T0205392

Big Data Management

The big data paradigm presents a number of challenges for university curricula on big data or data science related topics. On the one hand, new research, tools and technologies are currently being developed to harness the increasingly large quantities of data being generated within our society. On the other, big data curricula at universities are still based on the computer science knowledge systems established in the 1960s and 70s. The gap between the theories and applications is becoming larger, as a result of which current education programs cannot meet the industry's demands for big data talents.

This series aims to refresh and complement the theory and knowledge framework for data management and analytics, reflect the latest research and applications in big data, and highlight key computational tools and techniques currently in development. Its goal is to publish a broad range of textbooks, research monographs, and edited volumes that will:

– Present a systematic and comprehensive knowledge structure for big data and data science research and education
– Supply lectures on big data and data science education with timely and practical reference materials to be used in courses
– Provide introductory and advanced instructional and reference material for students and professionals in computational science and big data
– Familiarize researchers with the latest discoveries and resources they need to advance the field
– Offer assistance to interdisciplinary researchers and practitioners seeking to learn more about big data

The scope of the series includes, but is not limited to, titles in the areas of database management, data mining, data analytics, search engines, data integration, NLP, knowledge graphs, information retrieval, social networks, etc. Other relevant topics will also be considered.

More information about this series at http://www.springer.com/series/15869

Qi Xuan • Zhongyuan Ruan • Yong Min
Editors

Graph Data Mining

Algorithm, Security and Application

 Springer

Editors
Qi Xuan
Institute of Cyberspace Security
Zhejiang University of Technology
Hangzhou, China

Zhongyuan Ruan
Institute of Cyberspace Security
Zhejiang University of Technology
Hangzhou, China

Yong Min
Institute of Cyberspace Security
Zhejiang University of Technology
Hangzhou, China

ISSN 2522-0179   ISSN 2522-0187   (electronic)
Big Data Management
ISBN 978-981-16-2611-1   ISBN 978-981-16-2609-8   (eBook)
https://doi.org/10.1007/978-981-16-2609-8

This Springer imprint is published by the registered company Springer Nature Singapore Pte Ltd.
The registered company address is: 152 Beach Road, #21-01/04 Gateway East, Singapore 189721, Singapore

Preface

Things interact with each other, and our beautiful world emerges. Many real-world systems, natural or artificial, are more naturally expressed as graphs/networks, rather than coordinates in Euclidean space, to capture their topological properties. In biology, proteins regulate each other, and such physiological interactions make up the so-called interactomics of the organism; neurons connect to each other to process signals in brains, leading to the emergence of intelligence; and species depend on each other so as to form the complex ecosystem. Besides, modern transportation systems connect different cities across different countries, largely facilitating our travel and making the whole world a true global village. Nowadays, networking seems to be on the rise as we are entering the cyberspace. People are connected closely and share their viewpoints and personal interests through social media platforms such as Facebook, Wechat, Twitter, Weibo, and others. We can search our interests using search engines such as Google, Baidu, and Yahoo, the kernel of these systems being a huge network of webpages. We can also transfer money easily through e-banking or blockchain-based platforms such as Ethereum. Moreover, some powerful data mining or artificial intelligence technologies, such as knowledge graph and deep neural networks, are also networks! Although these networks facilitate information exchange among individuals, making our lives much easier than before, they may also facilitate the spread of virus and cause privacy leaks, e.g., particular kind of relationships could be inferred just based on individual ego social networks [1]. Therefore, it is necessary and urgent to develop methods to better understand the topological structure of these networks, so as to predict, and further influence, their evolution to a certain extent.

Fortunately, graph theory, as a branch of mathematics, has been well established since Euler's creative study on the Seven Bridges of Königsberg in 1736 [2]. In this big data era, more and more systems are described as networks, and the corresponding graph data capturing their structure are released for research. These graphs attract numerous researchers of different domains to contribute their talents to observe and further measure them from micro (node and link) and meso (motif and community) to macro (whole network) views by proposing a series of structural properties [3]. In industry, many famous search engines and recommender systems are essentially ranking nodes based on their structural significance in the corresponding networks, e.g., the well-known pagerank [4] and collaborative filtering algorithms [5]. On the other hand, in academia, Strogatz et al. [6] characterized small-world networks based on their short average distance together with large average clustering coefficient, while Barabási et al. [7] defined scale-free networks by power-law degree distribution. These studies triggered the development of complex network. Subsequently, various mathematical models are being proposed, with different kinds of dynamics, such as epidemic and synchronization, being carefully simulated and analyzed [8]. Quite recently, graph embedding technologies, such as deepwalk [9] and node2vec [10], were proposed to bridge between network space and Euclidean space; therefore, machine learning algorithms can be adopted to analyze graphs automatically. Soon, deep learning frameworks, such as graph convolutional network (GCN) [11, 12], were proposed to further facilitate the analysis of networks.

In this book, we mainly focus on typical supervised learning on graph data. In particular, the first three chapters introduce state-of-the-art graph data mining algorithms on node classification, link prediction, and graph classification, followed by a chapter that introduces the graph augmentation for further enhancement of those existing graph data mining algorithms. Chapters 5 and 6 analyze the vulnerability of these algorithms under adversarial attacks and the ways to increase their robustness, respectively. Note that we also analyze the vulnerability of community detection as unsupervised learning in Chap. 5 for a comprehensive review. Next, we provide brief reviews for node classification, link prediction, and graph classification, followed by the brief introduction of our own chapters.

Node classification helps to predict the labels of unknown nodes by attribute and structure information, which has been widely used in social network analysis. In some cases, nodes of the same category have similar topological properties, then we can use the basic structural features [13], including degree centrality, closeness centrality, betweenness centrality, eigenvector centrality, clustering coefficient, H-index, coreness, and PageRank of the target nodes, to distinguish them. We may either use one of them or a collection of them together with a machine learning algorithm to make the classification. Such models are relatively simple and interpretable, since we can easily know which feature is more significant in the model, and almost all of these handcraft features are of physical or social meaning. Although such simple models are feasible in certain cases, they may lose their effect in some others, especially when real-world networks may contain tens of thousands of nodes with irregular network topology. Graph embedding techniques are then developed to convert the graph data into a distinguishable vector representation while preserving the intrinsic graph properties. Common embedding techniques are based on random walk, matrix factorization, and neural network. And the typical graph embedding algorithms include: DeepWalk [9], which is based on skip-gram model [14] and uses the co-occurrence relationship information between nodes in the network to get the embedding vector of nodes; GraRep [15], which preserves the high-order proximity of graphs in the embedding space by matrix factorization; and LINE [16], which retains the first-order and second-order similarity between nodes, so as to present the local and global structure information of the network at the same time. Moreover, as the development of deep learning, a series of graph neural networks (GNN) were proposed to realize node classification end-to-end. Since Bruna [17] proposed the first spectral-based graph convolution neural network (GCN), which extends graph and convolution operation to spectral domain through Fourier transform, ChebNet [11] and CayleyNet [18] simplified the graph filter by Chebyshev and Cayley polynomials, respectively. Kipf and Welling [12] further simplified the operation by the first-order approximation of Chebyshev polynomials, which bridges the gap between spectral-based approaches and spatial-based approaches. These studies largely promote the development of spatial-based approaches and significantly improve the node classification performance. However, such deep models are highly nonlinear and complex, making them difficult to understand and increase the potential vulnerability.

In Chap. 1, Shu, Yu, Ruan, Zhang, and Xuan design a two-channel GNN to locate the source of an epidemic on a network, which can be considered as a typical node classification problem. Therein, node channel leverages the network structure to represent each node as an embedding vector, while edge channel transforms the original network into a line graph, and extracts feature vectors of nodes in line graph as representations of original edges. The features of two channels are then integrated to make the better estimation.

Link prediction aims to predict missed or future relationships based on currently observed connections [19], which has been widely adopted in social networks and biological networks. In social networks, link prediction is used to recommend probable friendships, leading to a more satisfactory user experience [20]. In biological networks, link prediction is adopted to predict previously unknown interactions between proteins, significantly reducing the costs of empirical approaches [21]. In addition, the data used to construct social and biological networks may contain inaccurate information, leading to spurious links [22, 23], which can also be identified by link prediction algorithms [24]. Similarly, link prediction can be realized just based on the predefined similarity between pairwise nodes. Such similarity indices may be either local or global [25]. For example, common neighbors, preferential attachment [7], Adamic-Adar, and resource allocation [26] are local indices since they only involve one-hop or two-hop neighbors of the target pair of nodes; Katz, rooted PageRank [27], and SimRank [28] are global indices, requiring to know the structure of entire network. A collection of these similarity indices may then be input to a machine learning model to get higher link prediction performance, by comparing with only one of them. This performance can be further improved by adopting graph embedding and GNN technologies, since the embedding vector of an edge can be easily obtained by combining those of the corresponding terminal nodes via a binary operator, i.e., Average, Hadamard, Weighted-L1, and Weighted-L2.

In Chap. 2, Zhang, Chen, and Xuan propose a hyper-substructure enhanced link predictor (HELP), as an end-to-end deep learning framework, for link prediction. HELP utilizes local topological structures from the neighborhood of the given node pairs, and learns features from hyper-substructure network for further exploiting higher-order structural information. Such method has relatively high efficiency and effectiveness, achieving state-of-the-art link prediction performance.

Graph classification focuses on classifying different networks through their structural differences. In chemistry, we may want to classify each compound as being either toxic or nontoxic [29] based on their structure. And this may be helpful in drug research and development [30], since the traditional discovery of new drugs is very expensive. In social networks, we may use a graph to capture individual behaviors, e.g., we can establish mobility networks [31] or focus shifting networks [32] for human beings based on their mobility or working traces, so that we can group them based on the structural properties of these networks. We can also classify teams based on their inside communication patterns. Again, we can simply use the graph statistics, such as degree distribution and average shortest path length, to classify different networks [33, 34]. Moreover, the entire graph can be processed to get the count of different graphlets or subgraphs, and one can also use these frequency statistics to produce a feature vector for graph classification [35, 36]. Another popular approach is defining graph kernel to measure the similarity between graphs, which can be plugged into a kernel machine. Numerous graph kernels have been proposed, including Shortest-Path kernel [37], Graphlet kernel [38], Random-Walk kernel [39], Weisfeiler-Lehman kernels [40], and Deep Graph Kernels [41]. Meanwhile, graph embedding techniques, e.g., Graph2Vec [42], and GNN techniques, e.g., DiffPool [43] and Graph Attention Network (GAT) [44], are also proposed for graph classification, achieving good performance.

In Chap. 3, Wang, Chen, Xie, Shan, Xuan, and Chen introduce the method to establish subgraph network (SGN), and further the sampling subgraph network (S^2GN) of higher scalability and diversity, by integrating different sampling strategies. They use both SGN and S^2GN to expand the structural feature space of the target networks, which, together with broad learning, significantly enhance the performance of a number of graph classification algorithms.

In Chap. 4, Zhou, Shen, Shan, Xuan, and Chen further introduce a novel iterative technique, namely M-Evolve, for graph data augmentation. M-Evolve includes subgraph augmentation, data filtration, and model retraining, which is applied for multi tasks including node classification, link prediction, and graph classification. Experiments validate that this method can help to overcome overfitting to certain extent, and significantly enhance a series of graph data mining algorithms.

Algorithm security mainly focuses on analyzing the vulnerability of graph data mining algorithms under certain kind of perturbation on network structure, and further proposing the corresponding defense strategies. Recent studies indicate that many artificial intelligence (AI) algorithms could be quite vulnerable under adversarial attacks, e.g., tiny perturbations can pollute an image and thereby reduce the classification performance significantly. Such adversarial attacks can also threaten graph data mining algorithms, i.e., their performance could be significantly reduced by slightly changing network structure. On the other hand, we can also design certain methods to detect and further defend against such attacks, so as to improve the robustness of graph data mining algorithms.

In Chap. 5, Shan, Zhu, Xie, Wang, Zhou, Zhou, and Xuan give a brief review on adversarial attacks for graph data mining. They briefly classify the existing attack methods, which could be heuristic, gradient, or reinforcement learning based. Then, for each graph mining task, one or two adversarial attack methods are introduced in detail. The experimental results show that most graph algorithms could be largely influenced by slightly modifying the structure or features of the graph.

In Chap. 6, Xu, Gan, Zhou, Wang, Chen, and Xuan on the other hand give a brief review on defense strategies against malicious attacks for graph data mining. They classify these strategies into five categories: adversarial training, graph purification, certifiable robustness, attention mechanism, and adversarial detection. Different kinds of strategies have different application scenarios. We hope this kind of research can remind researchers and engineers that the robustness of algorithms could be crucial in many real-world applications.

Applications are the driving force of the algorithm development. The rest six chapters introduce various applications of graph data mining algorithms in finance, social networks, transportation, communication, and epidemics. And we hope to draw the attention of the scientists and researchers in data mining, knowledge discovery, artificial intelligence, network science, as well as the related application domains. These chapters are briefly introduced as follows.

In Chap. 7, Xie, Zhou, Wang, Zhang, Sheng, Wu, and Xuan introduce a time-series snapshot network to model Ethereum transaction records as a spatial-temporal network, and define temporal biased walk to effectively embed accounts. Based on these, they use node classification and link prediction techniques to detect phishing and track transaction, respectively. Such study can help to better understand Ethereum transaction as a whole system from a network perspective.

In Chap. 8, Zhang, Xia, Li, Shen, Wang, and Xuan establish a friend network based on Yelp dataset and recommend friends through random forest and variational graph auto-encoder methods. The former takes multiple handcraft node similarity indices as input, while the latter automatically learns network structural features through a deep learning framework. They further construct a co-foraging network and recommend potential meal pals to users, validating the potential application of link prediction approaches to Yelp data analysis.

In Chap. 9, Xu and Dai introduce a graph convolutional recurrent neural network to predict traffic flow. They build a traffic network and then employ GCN model to learn the interactions between the roadways to capture the spatial dependence and use the long short-term memory (LSTM) neural network to learn traffic dynamic changes to capture temporal dependence. They make the prediction on a Hangzhou transportation network, validating the effectiveness of this method.

In Chap. 10, Qiu, Zhou, Cui, Chen, Zheng, and Xuan introduce circular limited penetrable visibility graph (CLPVG), mapping time series to graphs, which outperforms the traditional limited penetrable visibility graph (LPVG) method. Moreover, they also introduce the first end-to-end time series classification method based on GNN. The experimental results on several time-series datasets validate the outstanding performance of their methods.

In Chap. 11, Min, Zhou, Jiang, and Wu introduce the concept of social bots, including its definition, usage, and potential influence. They also introduce the related technologies to deploy social bots on social networks and summarize several methods to detect such social bots. In the future, various social bots could be increasingly deployed online as the development of AI technologies, which may pollute the Internet ecosystem and raise cyberspace security issues. Meanwhile, social bots, as artificial noise, may mislead graph data mining algorithms to a certain extent, which is worth to attract more attention from researchers in various areas.

Hangzhou, China   Qi Xuan

References

1. Backstrom, L., Kleinberg, J.: Romantic partnerships and the dispersion of social ties: a network analysis of relationship status on facebook. In: *Proceedings of the 17th ACM Conference on Computer Supported Cooperative Work and Social Computing*, pp. 831–841 (2014)
2. Biggs, N., Lloyd, E., Wilson, R.: Graph Theory, pp. 1736–1936 (Oxford University, Oxford, 1986)
3. Costa, L.D.F., Rodrigues, F.A., Travieso, G., Villas Boas, P.R.: Characterization of complex networks: a survey of measurements. Adv. Phys. **56**(1), 167–242 (2007)
4. Rogers, I.: The Google Pagerank Algorithm and how it Works (2002)
5. Sarwar, B., Karypis, G., Konstan, J., Riedl, J.: Item-based collaborative filtering recommendation algorithms. In: *Proceedings of the 10th International Conference on World Wide Web*, pp. 285–295 (2001)
6. Watts, D.J., Strogatz, S.H.: Collective dynamics of small-worldnetworks. Nature **393**(6684), 440–442 (1998)
7. Barabási, A.-L., Albert, R.: Emergence of scaling in random networks. Science **286**(5439), 509–512 (1999)
8. Boccaletti, S., Latora, V., Moreno, Y., Chavez, M., Hwang, D.-U.: Complex networks: Structure and dynamics. Phys. Rep. **424**(4–5), 175–308 (2006)
9. Perozzi, B., Al-Rfou, R., Skiena, S.: Deepwalk: Online learning of social representations. In: *Proceedings of the 20th ACM SIGKDD International Conference on Knowledge Discovery and Data Mining*, pp. 701–710 (2014)
10. A. Grover, J. Leskovec, node2vec: scalable feature learning for networks. In: *Proceedings of the 22nd ACM SIGKDD International Conference on Knowledge Discovery and Data Mining*, pp. 855–864 (2016)
11. M. Defferrard, X. Bresson, P. Vandergheynst, Convolutional neural networks on graphs with fast localized spectral filtering. Adv. Neural Inf. Process. Syst. **29**, 3844–3852 (2016)
12. Kipf, T.N., Welling, M.: Semi-supervised classification with graph convolutional networks. arXiv preprint arXiv:1609.02907 (2016)

13. Albert, R., Barabási, A.-L.: Statistical mechanics of complex networks. Rev. Mod. Phys. **74**(1), 47 (2002)

14. McCormick, C.: Word2vec Tutorial-the Skip-gram Model (2016)

15. Cao, S., Lu, W., Xu, Q.: Grarep: Learning graph representations with global structural information. In: *Proceedings of the 24th ACM International on Conference on Information and Knowledge Management*, pp. 891–900 (2015)

16. Tang, J., Qu, M., Wang, M., Zhang, M., Yan, J., Mei, Q.: Line: Large-scale information network embedding. In: *Proceedings of the 24th International Conference on World Wide Web*, pp. 1067–1077 (2015)

17. Bruna, J., Zaremba, W., Szlam, A., LeCun, Y.: Spectral networks and locally connected networks on graphs. arXiv preprint arXiv:1312.6203 (2013)

18. Levie, R., Monti, F., Bresson, X., Bronstein, M.M.: Cayleynets: graph convolutional neural networks with complex rational spectral filters. IEEE Trans. Signal Process. **67**(1), 97–109 (2018)

19. Liben-Nowell, D., Kleinberg, J.: The link-prediction problem for social networks. J. Am. Soc. Inf. Sci. Technol. **58**(7), 1019–1031 (2007)

20. Goyal, P., Ferrara, E.: Graph embedding techniques, applications, and performance: A survey. Knowledge-Based Syst. **151**, 78–94 (2018)

21. Martínez, V., Cano, C., Blanco, A.: Prophnet: a generic prioritization method through propagation of information. BMC Bioinf. **15**(S1), S5 (2014)

22. Von Mering, C., Krause, R., Snel, B., Cornell, M., Oliver, S.G., Fields, S., Bork, P.: Comparative assessment of large-scale data sets of protein–protein interactions. Nature **417**(6887), 399–403 (2002)

23. Butts, C.T.: Network inference, error, and informant (in) accuracy: a bayesian approach. Social Netw. **25**(2), 103–140 (2003)

24. Guimerà, R., Sales-Pardo, M.: Missing and spurious interactions and the reconstruction of complex networks. Proc. Natl. Acad. Sci. **106**(52), 22073–22078 (2009)

25. Lü, L., Zhou, T.: Link prediction in complex networks: A survey. Physica A: Stat. Mech. Appl. **390**(6), 1150–1170 (2011)

26. Zhou, T., Lü, L., Zhang, Y.-C.: Predicting missing links via local information. Eur. Phys. J. B **71**(4), 623–630 (2009)

27. Brin, S., Page, L.: Reprint of: the anatomy of a large-scale hypertextual web search engine. Comput. Netw. **56**(18), 3825–3833 (2012)

28. Jeh, G., Widom, J.: Simrank: a measure of structural-context similarity. In: *Proceedings of the Eighth ACM SIGKDD International Conference on Knowledge Discovery and Data Mining*, pp. 538–543 (2002)

29. Wale, N., Watson, I.A., Karypis, G.: Comparison of descriptor spaces for chemical compound retrieval and classification. Knowl. Inf. Syst. **14**(3), 347–375 (2008)

30. Jing, Y., Bian, Y., Hu, Z., Wang, L., Xie, X.-Q.S.: Deep learning for drug design: An artificial intelligence paradigm for drug discovery in the big data era. AAPS J. **20**(3), 58 (2018)

31. Song, C., Qu, Z., Blumm, N., Barabási, A.-L.: Limits of predictability in human mobility. Science **327**(5968), 1018–1021 (2010)

32. Xuan, Q., Okano, A., Devanbu, P., Filkov, V.: Focus-shifting patterns of oss developers and their congruence with call graphs. In: *Proceedings of the 22nd ACM SIGSOFT International Symposium on Foundations of Software Engineering*, pp. 401–412 (2014)

33. Li, G., Semerci, M., Yener, B., Zaki, M.J.: Effective graph classification based on topological and label attributes. Stat. Anal. Data Min. ASA Data Sci. J. **5**(4), 265–283 (2012)

34. X. Sun, S. Wandelt, Network similarity analysis of air navigation route systems. Transp. Res. Part E Logist. Transp. Rev. **70**, 416–434 (2014)

35. Benson, A.R., Gleich, D.F., Leskovec, J.: Higher-order organization of complex networks. Science **353**(6295), 163–166 (2016)

36. Nguyen, D., Luo, W., Nguyen, T.D., Venkatesh, S., Phung, D.: Learning graph representation via frequent subgraphs. In: *Proceedings of the 2018 SIAM International Conference on Data Mining*, pp. 306–314 (SIAM, New York, 2018)

37. Borgwardt, K.M., Kriegel, H.-P.: Shortest-path kernels on graphs. In: *Proceedings of the Fifth IEEE International Conference on Data Mining (ICDM'05)*, pp. 8–pp (IEEE, New York, 2005)

38. Shervashidze, N., Vishwanathan, S.V.N., Petri, T., Mehlhorn, K., Borgwardt, K.: Efficient graphlet kernels for large graph comparison. In: *Artificial Intelligence and Statistics*, pp. 488–495 (2009)

39. Kashima, H., Tsuda, K., Inokuchi, A.: Marginalized kernels between labeled graphs. In: *Proceedings of the 20th International Conference on Machine Learning (ICML-03)*, pp. 321–328 (2003)

40. Shervashidze, N., Schweitzer, P., van Leeuwen, E.J., Mehlhorn, K., Borgwardt, K.M.: Weisfeiler-lehman graph kernels. J. Mach. Learn. Res. **12**, 25392561 (2011)

41. Yanardag, P., Vishwanathan, S.V.N.: Deep graph kernels. In: *Proceedings of the 21th ACM SIGKDD International Conference on Knowledge Discovery and Data Mining*, pp. 1365–1374 (ACM, New York, 2015)

42. Narayanan, A., Chandramohan, M., Venkatesan, R., Chen, L., Liu, Y., Jaiswal, S.: graph2vec: learning distributed representations of graphs. In: *International Workshop on Mining and Learning with Graphs* (2017)

43. Ying, Z., You, J., Morris, C., Ren, X., Hamilton, W., Leskovec, J.: Hierarchical graph representation learning with differentiable pooling. In: *Advances in Neural Information Processing Systems*, pp. 4800–4810 (2018)

44. Veličković, P., Cucurull, G., Casanova, A., Romero, A., Liò, P., Bengio, Y.: Graph attention networks. In: *International Conference on Learning Representations* (2018)

Acknowledgment

We would like to thank Professor Guanrong Chen and Associate Professor Qing-peng Zhang at City University of Hong Kong, Associate Professor Jiajing Wu at Sun Yat-sen University, Associate Researcher Shilian Zheng at Science and Technology on Communication Information Security Control Laboratory, Professor Ye Wu at Beijing Normal University, Professor Jinyin Chen and Associate Professor Dongwei Xu at Zhejiang University of Technology, and all the other contributors. We also thank all the members in the IVSN Research Group, Zhejiang University of Technology, for the valuable discussions about the ideas and technical details presented in this book. This work was partially supported by the National Natural Science Foundation of China under Grant Nos. 61973273 and 61903334, and by the Zhejiang Provincial Natural Science Foundation of China under Grant Nos. LR19F030001, LY21F030017, and LGF21G010003.

Contents

Chapter 1
Information Source Estimation with Multi-Channel Graph Neural Network

Xincheng Shu, Bin Yu, Zhongyuan Ruan, Qingpeng Zhang, and Qi Xuan

Abstract The highly connectivity of modern society has led to the easy diffusion of harmful information, such as rumors, computer viruses, and infectious diseases, which brings great troubles to our society. Thus, estimating and timely quarantining the source of an epidemic or a rumor is highly critical. In this chapter, based on GNN, we propose a multi-channel graph neural network framework to efficiently locate information source. Different from previous methods, we turn this task into a learning problem and design two channels of feature inputs as a solution. Specifically, node channel leverages the network structure to represent each node as an embedding vector, which captures the important structural information of the node. Edge channel transforms the original network into a line graph, and extract feature vectors of nodes in line graph as representations of original edges. Finally, the features of two channels are aggregated together to estimate the probability of each node to be the source node. We evaluate our method on both synthetic and real-world networks. Extensive experiments demonstrate the effectiveness of our method on the task of identifying the information source.

1.1 Introduction

The spreading phenomena in our daily life is universal, such as the diffusion of innovations, news, computer viruses, and rumors. The propagation of harmful information or diseases, however, may bring great political and economic losses. For example, during the widely spread of COVID-19 virus around the world,

X. Shu · B. Yu · Z. Ruan (✉) · Q. Xuan
Institute of Cyberspace Security, Zhejiang University of Technology, Hangzhou, China

College of Information Engineering, Zhejiang University of Technology, Hangzhou, China
e-mail: zyruan@zjut.edu.cn

Q. Zhang
School of Data Science, City University of Hong Kong, Hong Kong SAR, China

© The Author(s), under exclusive license to Springer Nature Singapore Pte Ltd. 2021
Q. Xuan et al. (eds.), *Graph Data Mining*, Big Data Management,
https://doi.org/10.1007/978-981-16-2609-8_1

1

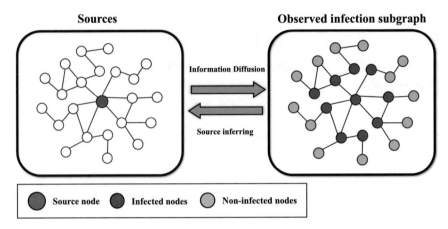

Fig. 1.1 Finding the information source aims to identify or detect by reversing information propagation in the network. The basic idea is to reversely estimate the location of the information source through the observed infection subgraph at a certain snapshot

both the disease and the related rumors on the Internet cause great troubles to us. Thus, how to control the spreading processes is of great importance in reality. To achieve this aim, one possible way is to model these spreading processes and understand the underlying mechanisms. So far, many epidemic and information diffusion models have been developed, such as the classical *Susceptible-Infection* (SI) model, the *Susceptible-Infected-Recovered* (SIR) model and the *Susceptible-Infected-Susceptible* (SIS) model etc. Based on these models, a large number of works have been focusing on the study of the range, the speed and the outbreak threshold of the information/epidemic spreading in networks [35, 38–40]. A reverse problem is how to locate the source nodes according to a given diffusion subgraph as shown in Fig. 1.1, which is extremely useful in controlling the information or epidemic spreading processes in practice.

Nevertheless, source inference is a difficult problem to solve due to a number of reasons. First, information diffusion process is highly dynamic and generally displays different patterns even starting from the same source. Second, in real cases, the underlying contagion mechanisms are usually unclear. While most of the existing source location methods assume a specific contagion model in advance. For example, Shah and Zaman first studied the problem on tree-like networks and assumed that the information diffusion process is characterized by the *Susceptible-Infected* (SI) model. Later works considered more realistic models such as the SIR and SIS models. However, some real spreading processes are much more complicated and thus can not be simply characterized by these models. Third, the existing mainstream methods are considered inadequate because of their high computational complexity and the prior knowledge (such as the subgraph structure and the propagation probability) they used may not correspond to the reality. For example, Lokhov et al. proposed the dynamic message passing (DMP) method to estimate the source, which uses all nodes in the network to calculate the marginal probability of a given node in a given state. However, DMP has low efficiency and is too time-consuming. Chang et al. [10] proposed a maximum a posteriori (MAP) estimator named as greedy search bound approximation (GSBA) to detect the information source with other methods (i.e., rumor center or Jordan center) as the prior. Both DMP and GSBA need to know the propagation probability between any two nodes for obtaining the maximum likelihood estimation. While in the real cases, the propagation probability is hard to exactly estimate. The aforementioned shortcomings greatly limit the application of the current algorithms in practical scenarios.

Consider an example as shown in Fig. 1.2, where the orange node indicates the source of the information, the red and blue nodes represent individuals who have received the information (infected). Note that the blue nodes are at the edge of the diffusion subgraph, connecting the non-infected nodes and the core infected nodes (red nodes). The aim is to identify the source node among the red and blue nodes given the diffusion subgraph. For a node which has received the information, it is logical to judge whether it is the source based on the states of its neighbor nodes. For example, node 2 has two non-infected neighbor nodes (i.e., node 3

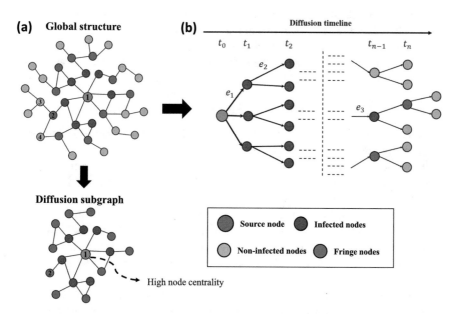

Fig. 1.2 The decomposition diagram of single information source detection problem. (**a**) After propagation model simulation, we can obtain the infection status of each node on the global structure topology. Moreover, nodes in the infected state can be extracted to form a diffusion subgraph and used to infer the true source. In general, the true source has a high node centrality. (**b**) Diffusion process is highly dynamic, and previous work do not consider higher-order network characteristics for inferring source nodes (i.e., edge importance). For example, obviously, the relationship between importance of edges e_1, e_2, e_3 is $e_1 > e_2 > e_3$

and 4). Therefore, node 2 is unlikely to be the source of the information, since intuitively, if node 2 is the source, its neighbor nodes are likely to be infected. Ali et al. [1] adopted the above idea and proposed a novel algorithm called Exoneration and Prominence based Age (EPA), which calculates the age of an infected node by considering the characteristics of both infected and non-infected neighbors. Moreover, from the diffusion subgraph (Fig. 1.2), we see that the source node is always in a core position, which can be reflected by various importance indicators of nodes. Recently, centrality-based methods have been widely used in the information source detection and achieve good performance. Moreover, the diffusion process has a character of temporality as shown in Fig. 1.2. In the diffusion subgraph, the importance of directed edges will increase with the distance from the source decreases. For example, obviously, the relationship between importance of edges e_1, e_2, e_3 is $e_1 > e_2 > e_3$. However, as far as we know, there is no combination of edge features and high-order structural features for information source detection.

In this chapter, we study the single information source detection problem (SISD) in which the underlying model of diffusion is assumed to be the heterogeneous SI model. This problem can be regarded as an end-to-end problem. The input is an undirected network topology and the features of each node (i.e., infection status and node centrality measures), and the output is the possibility of each node to be the source node. In recent years, graph neural network has been widely used in various tasks and practical scenarios on graph structured data because of its convincing performance and high interpretability. To solve SISD, Shah et al. [46] used graph neural networks (GNNs) to infer the information source without the knowledge of the underlying dynamics and its parameters. Yet [46] did not extract high-order structural features from diffusion subgraphs to infer information sources. Therefore, the major contributions of this chapter can be summarized as follows:

- First, based on GNN, we propose a multi-channel (node channel and edge channel) graph neural network framework (MCGNN) to efficiently locate the information source. The two channels leverage the network structure to extract the features of nodes and edges, respectively. By adopting feature fusion, the precision of the source detection is improved.
- Second, for each node in the diffusion subgraph, we construct the structural features and the prior knowledge features as the input of GNN to help estimate the source node. The prior knowledge features include the infection states of the nodes and the probability as the source for each node calculated by some previous approaches.
- Third, we conduct extensive experiments on synthetic networks and real-world datasets, demonstrating both the effectiveness and efficiency of our proposed model.

The rest of this chapter is organized as follows: Sect. 1.2 provides a brief review of related works. Section 1.3 introduces some preliminaries of single information source detection. Section 1.4 shows the architecture and details of our proposed model MCGNN. Experimental results are reported in Sect. 1.5. Finally, Sect. 1.6 concludes this chapter.

1.2 Related Work

Information source detection has attracted great attention and has been extensively studied over the past decade. In this section, we mainly review the relevant research from the following three categories: (1) information diffusion modeling. (2) information source detection and (3) graph neural network.

1.2.1 Information Diffusion Modeling

Modeling the spreading processes is of great significance for controlling the epidemics or rumors in social networks. It has attracted researchers from different fields like computer science, epidemiology, sociology and physics [52]. A large number of models have been proposed, such as the SI model, SIS model, SIR model, and linear threshold model (LT). In the SI model, individuals have two states: *Suspected, Infected*. A susceptible individual will be infected with a certain probability if he/she encounters an infected individual, and the infected agents will keep their state until the end of the dynamics. In the SIS model, however, the infected nodes will recover to the susceptible state with a certain probability, thereafter, participating in the infection process again. Differently, in the SIR model, the infected nodes will turn into removed state, and never be infected or infect others again. LT model assumes that a node can be infected only if the fraction of infected neighbors of the node is greater than a threshold. In this chapter, we use the heterogeneous SI model to generate the diffusion subgraph, in which the infection probabilities on different links are heterogeneous.

1.2.2 Information Source Detection

The problem studied in this chapter can be regarded as the reverse reasoning process of information diffusion modeling. Various methods have been proposed to identify a single source under the SI model. For example, Shah and Zaman [44, 45] studied the single source detection problem with the popular SI model for the first time and constructed an estimator for the information source, which is termed as Rumor Centrality (RC). For each node, RC represents the number of permuted propagation paths and their corresponding probabilities. However, the authors only considered a simple case that the underlying network is unweighted. Luo et al. [31] studied the problem, and proposed an algorithm with quadratic complexity to estimate the actual number and identities of the infection sources. Meanwhile, Luo et al. [32] also considered the problem of estimating an infection source with Limited Observations. Dong et al. [13] constructed a maximum a posteriori (MAP) estimator to identify the rumor source with different settings of the prior. To handle the analysis of the MAP estimator, they also developed a key concept of *local rumor center*, which originates from RC. Wang et al. [48] addressed the problem of rumor source detection with multiple observations under the SI model. For tree networks, the authors found that multiple independent observations can dramatically improve the detection probability. Jain et al. [22] proposed a heuristic method based on the hitting time statistics of a surrogate random walk process that can be used to approximate the maximum likelihood estimator of the rumor source. Chang et al. [10] proposed a maximum a posteriori (MAP) estimator named as greedy search bound approximation (GSBA) to detect the information source with other methods as the prior. Choi et al. [11] studied the problem by querying the individuals under the SI model, given a sample snapshot of the information diffusion graph. They proposed two practical estimation algorithms, each of non-adaptive (NA) and adaptive (AD) types, and for each algorithm, the authors quantitatively analyzed the budget which ensures $1 - \delta$ detection accuracy.

Besides, researchers also studied the problem under more general epidemic models, such as SIR and SIS models. Zhu et al. [56] developed a sample path based approach to detect the information source in tree graphs under the SIR model and proposed the *Reverse Infection* (RI) algorithm to find the source in general graphs which is proved to be its Jordan center (JC) [24]. The estimator of the information source is chosen to be the root node associated with the sample path according to the observed diffusion subgraph. Luo et al. [30] derived an estimator based on estimating the most likely infection source associated with the most likely infection path, assuming that the infection process follows the SIS model. Moreover, considering the SIR model, Lokhov et al. [29] introduced a new algorithm based on the dynamic message passing (DMP) equations for the estimation of the source by given the network topology and the snapshot of some nodes at a certain time. It used a mean-field-like approximation to compute the probability of the observed snapshot as a product of the marginal probabilities, which is the basis of node ranking. Whereas DMP method is too time-consuming to be applied to real scenarios, and the propagation probability which is used to calculate the marginal probabilities is extremely hard to capture. Unlike the most existing methods, Altarelli et al. [4] performed Bayesian inference for the information source detection under the SIR model. They derived the Belief Propagation (BP) equations which can accurately calculate the posterior distribution of each node state by time evolution. Furthermore, they also generalized the problem with noisy observations [3].

In addition, many researchers focused on the problem of detecting multiple information sources. Prakash et al. [36] proposed to employ the Minimum Description Length (MDL) principle to identify the best set of seed nodes and virus propagation ripple, which describes the infected graph most succinctly. They proposed a highly efficient algorithm, namely Netsleuth, to identify likely sets of seed nodes given a snapshot. Given these seed nodes, the authors showed that Netsleuth can optimize the virus propagation ripple in a principled way by maximizing likelihood. Fioriti et al. [16] introduced a dynamic age method to identify multiple diffusion sources in general networks. They claimed that the oldest nodes, which are associated to those with largest eigenvalues of the adjacency matrix, are the sources of the diffusion. Zhu et al. [57] proposed a new source localization algorithm, named Optimal-Jordan-Cover (OJC). The algorithm first extracts a subgraph using a candidate selection algorithm which selects source candidates based on the number of observed infected nodes in their neighborhoods. Considering the heterogeneous SIR model in the ER random graph, they proved that OJC can locate all sources with probability one asymptotically with partial observations. Jiang et al. [23] proposed a novel method, i.e., K-center method, to identify multiple diffusion sources in general networks, which can address the question of how many sources there are and where the diffusion emerges. Wang et al. [49] proposed Label Propagation based Source Identification (LPSI) algorithm, which exploits the idea of source prominence to find multiple sources without any information of the underlying propagation model using label propagation mechanism. Ali et al. [1] proposed a novel algorithm called Exoneration and Prominence based Age (EPA), which calculates the age of an infected node by considering the characteristics of both infected and non-infected neighbors.

1.2.3   Graph Neural Network

Graph neural networks (GNN) [8, 18, 51, 53] have received considerable attention from various fields, such as social science [20, 27], physics [7, 41], biology [17], knowledge graphs [19, 34] and many other research areas [26].

The concept of GNN was first proposed in [42], which extends the existing neural networks for processing the data represented in graph domains. Kipf and Welling [27] proposed a spectral approach, called GCN, which employs a localized first-order approximation of graph convolutions. More recently, Veličković et al. [47] proposed graph attention networks (GATs) to aggregate localized neighbor information based on attention mechanism. For the problem of information source detection, Dong et al. [14] proposed a deep learning based model, namely GCNSI, to locate multiple rumor sources without prior knowledge of underlying propagation model. Futhermore, Shah et al. [46] revisited this problem using graph neural networks (GNNs) to learn Patient Zero (P0). They established a theoretical limit for the identification of P0 in a class of epidemic models. Our work is also based on the graph convolution network architecture, we consider the node importance features and the prior knowledge of each infected node in the diffusion subgraph, and set up multiple channels to extend this architecture to learn the information source from multiple perspectives.

Table 1.1 Notation summarization

Symbols	Descriptions
$G = (V, E, W)$	An undirected and weighted social network.
G_I	An diffusion subgraph of G.
η_{ij}	Probability of propagation between node i and j.
V	Set of vertexes in G.
E	Set of edges in G.
$Y = (Y_1, \cdots, Y_{|V|})^T$	Infection state of social network
$Y_i \to \{1, 0\}$	Infection state of a single node i.
s	Predictable information source.
s^*	Ground-truth of information source.
λ	Weight of L2 regularization.
y'	Output of MCGNN.
y	Ground-truth vector of information sources.

1.3 Preliminaries

In this section, we introduce the basic knowledge and relevant techniques that are necessary for understanding MCGNN. Frequently used notations are summarized in Table 1.1.

1.3.1 Problem Definition

Let's consider an undirected and weighted social network $G = (V, E, W)$, where V is the node set, E is the edge set, and $W = [\eta_{ij}]$ ($\eta_{ij} \in [0, 1]$) is the information propagation probability from node i to j. Let $Y = (Y_1, \cdots, Y_{|V|})^T$ be the infection state of all nodes in G. The state of each node $Y_i \in \{1, 0\}$ denotes the infection state of node $v_i \in V$, in which $Y_i = 1$ and $Y_i = 0$ indicate that v_i is infected and uninfected, respectively. In this chapter, we assume the source consists of a single node s, and apply the heterogeneous SI model to characterize the diffusion process.

The heterogeneous SI model is a variant of the popular SIR model [25]. It assumes that each node has two possible states: (1) susceptible and (2) infected. Once a node i is infected (or receives the information), it will remain in the state until the end of the dynamics. Meanwhile, the infected node i will spread the information to its susceptible neighboring node j with probability η_{ij}. The infections along different edges are supposed to be independent. After the information spreads on the network for some time, there will be a group of infected nodes, denoted by V_I, which includes the source node s. These nodes and their interconnections E_I can span a diffusion subgraph $G_I(V_I, E_I)$ of $G(V, E)$, which are referred to as G_I and G, respectively. G_I is connected because every susceptible node can only be infected by its neighbors.

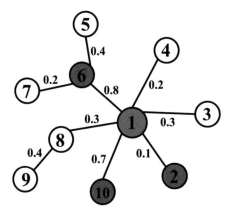

Fig. 1.3 Diffusion subgraph under the SI model, where red and brown nodes are infected and others are susceptible. Numbers on edges are the information propagation probability between two adjacent nodes

As an example, Fig. 1.3 shows a graph in which nodes 1, 2, 6, and 10 are infected (forming a diffusion subgraph), and other nodes are susceptible. In our experimental setup, the data that we can observe include only the graph structure and a snapshot of the spreading process (i.e., the state of each node at a given time is known), while the time for each infection event and the propagation probability between each node pair are unknown. The question is how to locate the source node based on these information. To solve this problem, let's start by introducing the following definitions:

Definition 1 (Global Graph) Global graph $G(V, E, W)$ is the original network over which the information spreads, where V is the set of nodes and E is the set of edges. $W = [\eta_{ij}]$ is the information propagation probability between nodes i and j.

Definition 2 (Diffusion Subgraph) Diffusion Subgraph $G_I(V_I, E_I)$ is a connected subgraph of G, consisting of the infected nodes $V_I \subseteq V$, and the corresponding edges $E_I \subseteq E$.

Definition 3 (Source) A source s^* is the node from which the information starts on the global graph G.

Definition 4 (Single Information Source Detection) Given a diffusion subgraph G_I and its corresponding global graph G, the problem undertaken in this study is to identify the infection source s^* in G_I, assuming that G_I has been infected by a single source under the heterogeneous SI diffusion model.

Definition 5 (Line Graph) Given a network $G(V, E)$, the line graph, denoted by $G^* = L(G)$, is a mapping from G to $G^*(V^*, E^*)$, in which $V^* = \{v_1^*, v_2^*, \cdots, v_{|E|}^*\}$ and $E^* \subseteq (V^* \times V^*)$ represent the set of nodes and links, respectively. The mapping process is as follows: As shown in Fig. 1.4, if two nodes v_i and v_j are connected in the original network, the link between them

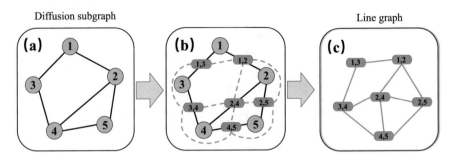

Fig. 1.4 The process of building line graph from a given network: (**a**) the original network (i.e., diffusion subgraph in our task), (**b**) extracting lines as nodes and establishing connections among these lines, and (**c**) forming line graph

is transformed as a node, and two nodes in the line graph are connected if the corresponding links share the same terminal node in the original network.

1.4 Multi-Channel Graph Neural Network

In this section, based on GNN, we introduce a multi-channel graph neural network framework to efficiently locate the information source. Different from the previous methods, we turn this task into a learning problem and design two channels (i.e., node channel and edge channel) of feature inputs as a solution. Specifically, node channel leverages the network structure to represent each node as an embedding vector, which captures the important structural information of the node. Edge channel transforms the original network into a line graph, and extract feature vector of the line graph as representations of original edges. Finally, the features of two channels are aggregated together to estimate the probability of each node to be the source node. In terms of that, our framework MCGNN combining node level vector with high-order edge features can enhance the performance of information diffusion source reasoning.

1.4.1 Feature Indices of Input

In this work, we extract two sets of features: node features (e.g., structural features or priori knowledge) on diffusion subgraph and edge features on line graph. Node features in diffusion subgraph are often used for information source detection [2, 12]. However, in the previous studies of single information source detection, the importance of edges is largely ignored. We here try to fill this gap. To proceed, we first convert the original global graph to line graph, then calculate the node centrality indices of the line graph to directly define the edge importance in the original network. In addition, the features of uninfected nodes in the global graph are set to zero, to distinguish from the infected nodes.

1.4.1.1 Structural Features

In particular, the structural features in diffusion graph and the line graph include:

- Degree Centrality (D). It is defined as

$$DC_i = \frac{k_i}{N-1}, \tag{1.1}$$

where k_i is the degree of node i and N is the total number of nodes in the corresponding graph (i.e., diffusion subgraph or line graph).
- Closeness Centrality (CC). It is defined as

$$CC_i = \frac{N}{\sum_{j=1}^{N} d_{ij}}, \tag{1.2}$$

where d_{ij} denotes the shortest path length between nodes i and j in the corresponding graph. The shorter the distances between node i and the rest nodes are, the more central the node i is, and thus the larger CC_i index is.
- Betweenness Centrality (BC). It is defined as

$$BC_i = \sum_{s \neq i \neq t} \frac{n_{st}^i}{g_{st}}, \tag{1.3}$$

where g_{st} is the total number of shortest paths between nodes s and t in the corresponding graph, and n_{st}^i represents the number of shortest paths between nodes s and t that pass through node i.
- Eigenvector Centrality (EC). Eigenvector Centrality is also known as eigencentrality. It is defined as

$$EC_i = \alpha \sum_{j=1}^{N} a_{ij} EC_j, \tag{1.4}$$

where a_{ij} is the element of the adjacency matrix of the corresponding graph, i.e., $a_{ij} = 1$ if nodes i and j are connected and $a_{ij} = 0$ otherwise, and α should be less than the reciprocal of the maximum eigenvalue of the adjacency matrix.

- Clustering Coefficient (C). In this study, we consider the local clustering coefficient. It is defined as

$$C_i = \frac{2L_i}{k_i(k_i - 1)},\qquad(1.5)$$

  where L_i is the number of links between the k_i neighbors of node i.
- H-index (H). H-index is a popular metric which is used to measure both the productivity and citation impact of a scholar or scientist [21]. Sorting the degree of node $i's$ neighbors by decreasing order, the H-index can be calculated as following:

$$H_i = \max_{j \in \mathcal{N}(i)} \min(k_i, j),\qquad(1.6)$$

  where $\mathcal{N}(i)$ denotes the set of neighbors of node i.
- Coreness (CO). The coreness is defined based on k-core. The k-core of a network is defined as the maximal subnetwork in which each node has at least degree k [6, 43]. If a node belongs to k-core but not $(k + 1)$-core, it has coreness k [9].
- PageRank (PR). PageRank is a popular way of measuring the importance of website pages [33]. The underlying assumption is that more important webpages tend to have more links from other webpages. Its iterative formula is defined as

$$PR_i(t) = (1 - c) \sum_{j=1}^{N} a_{ji} \frac{PR_j(t - 1)}{k_j} + \frac{c}{N},\qquad(1.7)$$

where c is a free parameter between zero and one. In this study, we set $c = 0.15$.

1.4.1.2 Prior Knowledge Features

There have been many methods to study the information source inferring problem, such as Distance Centrality, Jordan Centrality, Rumor Centrality, LPSI [49] and EPA [1]. We here combine these methods with GNN to improve the detection precision. In particular, based on these methods, the importance value of each infected node in the diffusion subgraph will be calculated and input into our MCGNN framework as the features of each node. In the following, we will introduce these methods briefly.

- The most basic measure for node i being the information source is the distance centrality $D(i)$, which is defined as

$$D(i) = \sum_{j \in G_I} dis(i, j),\qquad(1.8)$$

where $dis(i, j)$ is the shortest path length between node i and node j in the graph G_I.

- Jordan centrality $J(i)$ is defined as the maximum distance from node i to other infected nodes [55]:

$$J(i) = \max_{j \in G_I} dis(i, j), \tag{1.9}$$

The node i with minimal $J(i)$ is known as a "Jordan center" of G_I.

- Rumor Centrality was proposed to detect the source with maximum likelihood estimation under the homogeneous SI model which assumes the propagation probability along each edge is equal. RC is defined as the number of permitted permutations starting with i:

$$RC(i) = \prod_{j \in G_I} \frac{N!}{T_j^i}, \tag{1.10}$$

where j is a node in G_I and T_j^i is the number of nodes in the subtree rooted at j with i being the source.

- Label Propagation based Source Identification (LPSI) was first proposed in [49], which was inspired by a label propagation based semi-supervised learning method [54]. In LPSI, the iteration formulation and the convergent state are defined as follows:

$$\mathscr{G}_i^{t+1} = \alpha \sum_{j \in \mathscr{N}(i)} S_{ij} \mathscr{G}_j^t + (1 - \alpha) Y_i, \tag{1.11}$$

$$\mathscr{G}^* = (1 - \alpha)(I - \alpha S)^{-1} Y. \tag{1.12}$$

In Eq. (1.11), $\alpha \in (0, 1)$ is the parameter used to control the influence from the neighbors to node i. G_i^t is the infection state of node i at time t. S_{ij} is the (i, j)-th element of the regularized Laplace matrix S of G_I. Y_i is the given infection state of node i. In Eq. (1.12), $\mathscr{G}^*$ is the convergence state of the network. I is the identity matrix. $S = D^{-1/2} W D^{-1/2}$ is the regularized Laplace matrix of G_I, where D is a diagonal matrix with its (i, i)-th element equaling to the sum of the i-th row of W, and W is the adjacency matrix of graph G_I.

- Exoneration and Prominence based Age (EPA) was proposed by Ali et al. [1], which exploits the concepts of exoneration effect and local prominence. Starting from any node i, the authors applied the BFS (Breath-First Search) algorithm to traverse the diffusion subgraph. Since the nodes generated by BFS in iteration l are l hops away from i, these nodes are considered to belong to level l. For a node j at level l, the prominence is defined as

$$P_j^l = (\frac{I_j}{O_j}) / (\frac{1}{1 + \ln O_j}), \tag{1.13}$$

where I_j and O_j are the corresponding degree of node j in the diffusion subgraph and in the global graph, respectively. Therefore, the age of node i will be the sum of the prominence of each level starting from level 0–$r-1$.

$$A(i) = \frac{\sum_{l=0}^{r-1} \sum_{v \in V_l} P_j^l}{ECC(i)}, \tag{1.14}$$

where r denotes the radius of the subgraph G_I, and $ECC(i)$ is the eccentricity of node i in the subgraph G_I.

1.4.2 Graph Convolutional Networks

Graph Convolutional Network (GCN) [27] is an extension of Convolutional Neural Network (CNN) on graph data, which generates local permutation-invariant aggregation on the neighborhood of a node in a graph such that the features of a graph can be efficiently captured. The GCN model is built by stacking multiple GCN layers. The input to each GCN layer is a vertex feature matrix, $H \in R^{N \times F}$, where N is the number of vertices, and F is the number of features. Each row of H, denoted by h_i^T, is associated with a vertex. Generally speaking, the essence of the GCN layer is the nonlinear transformation that outputs $H' \in R^{N \times f}$:

$$H' = GCN(H) = \sigma(A(G)HW^T + b), \tag{1.15}$$

where $W \in R^{f \times F}$, $b \in R^f$ are model parameters, σ is a non-linear activation function, $A(G)$ is an $n \times n$ matrix that captures structural information of graph G. GCN instantiates $A(G)$ to be a static matrix closely related to the normalized graph Laplacian matrix:

$$A(G) = D^{-1/2} \widetilde{A} D^{-1/2}. \tag{1.16}$$

where $\widetilde{A} = A + I$ is the adjacency matrix of G, I is the identity matrix, and $D_{ii} = \sum_j \widetilde{A}_{ij}$ represents the degree matrix of G.

1.4.3 Architecture of MCGNN

By applying the GCN framework, we here study the single information source detection (SISD) problem. The input is a number of features of the nodes in a given network, and the output is the possible information sources. According to the nature of SISD, it can be considered as a variant of the multi-label classification problem that aims to assign a two-valued label $(0, 1)$ to each node in G. In our model, we consider both the node importance and edge importance. The latter one has been largely ignored in the previous studies, while may be of great importance in detecting the source. To test our assumption, we extract the feature of each node in the line graph as the extended feature of the original diffusion subgraph. The results show that extracting both the node and edge features can effectively improve the accuracy of information source detection. As GCN can effectively capture the features of vertex domain and spectral domain for a graph, it is therefore proper to adopt GCN for expressing complicated neighborhood information.

As shown in Fig. 1.5, the architecture of our model (called MCGNN) can be divided into two main channels, i.e., node-level channel and edge-level channel. First, the set of training samples are generated by the heterogeneous SI model under global graph G. Second, we construct the features of each node in the simulated diffusion subgraph, including structural features and rumor-related features, which are input to GCN layers as the feature matrix W. Meanwhile, we sample the 2-hop ego network of each node in the diffusion subgraph, and extract the global structure features of the corresponding line graph as the importance of edges. Finally, the output matrix of GCN and the features matrix of line graph are aggregated as the final matrix, which include node-level part and edge-level part. In addition, we adopt a dense (full connected) layer and sigmoid function to transform the output matrix into a probability vector, where each value represents the probability that the corresponding node is an information source node. The prediction output of MCGNN is denoted by y'.

1.4.4   Loss Function

We denote a training sample for MCGNN as (x, y), where the input is x and the output is y. For a given input x, the corresponding output y is the true label of each node (regarding whether it is the information source or not). As mentioned before, SISD is a variant of the multi-label classification problem and we need to predict whether each node is an information source at the same time. Therefore, we adopt a Sigmoid cross entropy loss as the loss function. In addition, we use L2 regularization in the loss function to reduce overfitting. The loss function is described as follows:

$$\mathscr{L}(y', y) = -\log \sigma(y') \times y - \log(1 - \sigma(y')) \times (1 - y) + \lambda \|w\|_2, \qquad (1.17)$$

where y' is the output of MCGNN with the input x, y is the true label and σ is the Sigmoid function. w represents all of the weights in MCGNN and $\|w\|_2$ is the L2 regularization item, companied by a weight coefficient λ.

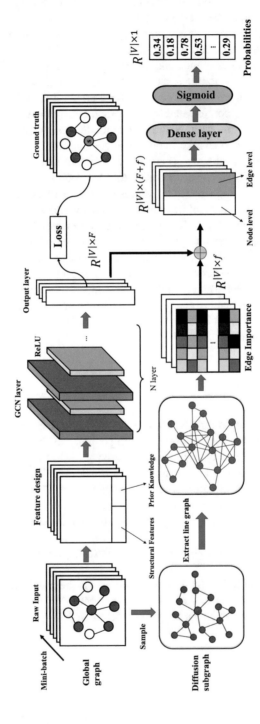

Fig. 1.5 Architecture of MCGNN. We improve the accuracy of estimating information source nodes by extracting the edge importance features of diffusion subgraph

1.5 Experiment

In this section, we first introduce the experimental settings, including the datasets, baselines and evaluation metrics. Then, we explore the detection effect of our method on both small and large scale diffusion subgraphs, and compare our method with baselines on different synthetic and real-world networks.

1.5.1 Datasets and Experimental Setup

We evaluate the performance of our proposed method for single information source detection on six different networks, including three synthetic networks (i.e., ER random [15], BA scale-free [5] and 4-regular network) and real-world networks (i.e., Email-univ [37], Facebook [28], US Power Grid (USPG) [50]). The basic statistics of these networks are presented in Table 1.2. In order to analyze the universality of our proposed method, we consider two different diffusion subgraph sizes (small scale and large scale). The number of nodes in the small scale experiments ranges from 20 to 60 with an interval of 5, and the number of nodes in the large scale experiments ranges from 400 to 600 with an interval of 100 (see the last two columns in Table 1.2).

1.5.2 Baselines and Evaluation Metrics

In order to verify the performance of our method MCGNN, we compare it with the following methods. In the experiments, we use the same settings, for example, keeping the number of independent experiments the same, and applying the SI model for all experiments.

- **Distance Centrality (DC):** DC selects the infected node with the smallest distance centrality as the source after ranking. Distance centrality is the sum of the shortest distances from a node to others [44].

Table 1.2 Dataset statistics

Netowrk	# Node	# Edge	# Degree	Small scale	Large scale
BA scale-free	1000	2991	5.982	20–60	400–600
ER random	1000	4000	8.0	20–60	400–600
4-regular	1000	2000	4.0	20–60	400–600
Email-univ	1133	5451	9.622	20–60	400–600
Facebook	4039	88,234	43.961	20–60	400–600
US Power Grid (USPG)	4941	6594	2.669	20–60	400–600

- **Jordan Centrality (JC):** Jordan center is the node which minimizes the maximum distance to others. JC selects the Jordan center as the source [24].
- **Rumor Centrality (RC):** Rumor center is the node with the maximal rumor centrality, which is selected as the source. RC is defined as the number of permitted permutations starting with one node [44].
- **Reverse Infection (RI):** RI is a low complexity algorithm to find the sample path based estimator in general graphs. It allows each infected node to broadcast a message containing its identity to its neighbors. Each node, after receiving messages from its neighbors, will record the arriving time and repeat the above process. The node with the minimum sum of arriving times will be selected as the source [56].
- **Dynamic Message Passing (DMP):** The algorithm uses DMP equations to estimate the source by given the network topology and the snapshot of some nodes at a certain time. It uses a mean-field-like approximation to compute the probability of the observed snapshot as a product of the marginal probabilities, and selects the node with the maximum probability, as the source [29].
- **Dynamic Importance (DI):** DI selects an infected node, which has the maximal reduction of the largest eigenvalue of the adjacent matrix after being removed from the network as the source [16].
- **Greedy Search Bound Approximation (GSBA):** GSBA is an approximate maximum a posteriori (MAP) estimator. It exploits a strategy of greedy search to find a surrogate upper bound of the likelihood of a permitted permutation [10]. In this chapter, we select Rumor Centrality as its prior knowledge.
- **Exoneration and Prominence based Age (EPA):** EPA calculates the age of an infected node by considering its prominence in terms of its both infected and non-infected neighbors. The oldest node is selected as the source [1].

We repeat 1000 independent experiments on the small-scale diffusion subgraphs to achieve the statistically meaningful results. While for large-scale diffusion subgraphs, some algorithm (e.g., DMP) is too time-consuming, thus we repeat 100 independent experiments on them. For MCGNN, we generate 1000 diffusion subgraphs for each network, then use 90–10 train-test split and 10-fold cross validation. Finally, we report the average result for all trials.

To evaluate our proposed method quantitatively, we use the following three widely adopted performance metrics:

- **Precision (*Prec.*):** Precision denotes the proportion of times that the information source is correctly detected in Q independent repeated experiments. It is defined as

$$P = \frac{Q_T}{Q},$$ (1.18)

where Q_T is the number of detection experiments in which the information source is located correctly.

- **Error Distance (*ED*):** Error Distance is the shortest topological distance between the ground truth s and the estimated source s^* in all trials. Formally, ED is defined as

$$ED = dis(s, s^*), \qquad (1.19)$$

where $dis(s, s^*)$ is the distance between the actual source s and the estimated source s^* in each detection experiment.

- **Normalized Ranking (*NR*):** For each method, we sort the probability or centrality value of each infection node as the source in descending order, and then normalize the ranking of the true source. NR is defined as

$$NR = \frac{R(s^*) - 1}{N_I}, \qquad (1.20)$$

where N_I is the size of the diffusion subgraph, and $R(s^*)$ denotes the ranking of the ground truth s^*. Clearly, smaller NR means the estimation result is better.

1.5.3 Results on the Synthetic Networks

In this subsection, we compare our method, namely MCGNN, with other baseline methods, i.e., DC, JC, RC, RI, DMP, DI, GSBA, EPA and GCN. Among these baselines, DC, JC RC, RI, DI, EPA, GCN are designed based on the topological characteristics of the diffusion subgraph, while DMP and GSBA are based on the probabilistic likelihood estimation, which require the prior knowledge, i.e., the propagation probability on each edge in advance. In the following, we will evaluate the source detection performance of each baseline and our method on the BA scale-free networks, ER random networks and 4-regular networks, respectively. For each kind of networks, we perform experiments on both small and large diffusion subgraphs, which are generated under the heterogeneous SI model.

First, we perform extensive experiments on small diffusion subgraphs of which the size ranges from 20 to 60 with an interval of 5. Figure 1.6 shows the precision, error distance and normalized ranking across all the three synthetic networks. In general, our method achieves better performance in most cases under all the three metrics.

As shown in Fig. 1.6a–c, we see that for BA scale-free networks, DI and RI algorithms perform poorly, which, for example, only achieve less than 20 % average precision. The probabilistic likelihood estimation based methods (GSBA and DMP) perform much better due to the prior knowledge. Other methods like DC, JC, RC and EPA, which estimate the source based on the topological structure, achieve around 40 % precision, worse than GSBA and DMP. The reason is clear: For example, if the source node in BA scale-free networks has a very small degree, the centrality based methods may probably fail, meaning that it is usually not reliable to estimate the source based on single topological measures in heterogeneous networks. Our method, which extracts the features of both nodes and edges can improve the estimation results and perform the best.

The results of ER random networks are reported in Fig. 1.6d–f, where we see that all the methods can achieve good performance. For example, compared to the case of scale free networks, the average precision increases for each method. Specifically, the precision of MCGNN almost reaches to 100% (Fig. 1.6d). Furthermore, the overall error distance decreases. Even for DI, which performs worst, the error distance is below 0.7 (Fig. 1.6e), indicating that the shortest path length between the estimated and the real source is less than 1. Finally, the normalized ranking also decreases significantly. The key point here is that the ER random network can be approximately regarded as a regular tree [56], which is a network structure with

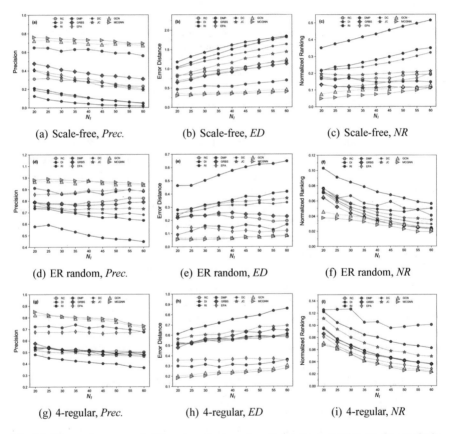

(a) Scale-free, *Prec.*   (b) Scale-free, *ED*   (c) Scale-free, *NR*

(d) ER random, *Prec.*   (e) ER random, *ED*   (f) ER random, *NR*

(g) 4-regular, *Prec.*   (h) 4-regular, *ED*   (i) 4-regular, *NR*

Fig. 1.6 The performance of different methods under all the three metrics on the synthetic networks. We test all the algorithms 1000 times and report the results on average. Synthetic networks include (**a**)–(**c**) BA scale-free network, (**d**)–(**f**) ER random network and (**g**)–(**i**) 4-regular network

a high inference precision according to the centrality based and probability based methods.

Finally, we repeat the above experiments on 4-regular networks. As shown in Fig. 1.6g–i, we find that MCGNN, GCN and EPA algorithms perform well on all the three metrics—the average precision is high, the error distance is small, and the normalized ranking is at top, meaning that these methods are relatively robust. While DMP performs abnormally, corresponding to high values of normalized ranking. In addition, the results for DI, DC, JC and RC are unsatisfactory (Fig. 1.6g, h), since for 4-regular networks where each node has the same degree, it is difficult for the methods which only extract single topological property to estimate the source accurately.

Next, we also conduct experiments on large diffusion subgraphs which at least 400 nodes have been infected. Here, we consider three different sizes of subgraphs, $N_I = \{400, 500, 600\}$. The results are shown in Table 1.3. In general, our method achieves better performance in most cases under all the three metrics, although there are some exceptions. For example, in ER random networks, RC performs the best in *Normalized Ranking*. Moreover, in 4-regular networks, DMP performs the best in *Precision* and *Error Distance* when the diffusion subgraph is large. These results suggest that the methods which combine multi-features of the underlying network can more accurately detect the source node, while it is still a challenging task on large-scale graphs.

To summarize, our method always achieves the best performance on the three synthetic networks, especially with respect to *Precision* and *Error Distance*. The reason is that MCGNN can extract features from different levels, i.e., node level and edge level. The above results suggest that our method has great advantage, which can infer the source node without the propagation probability, and have a good performance both on small and large diffusion subgraphs.

1.5.4 Results on the Real-World Networks

Next, we will show the experimental results on the real-world networks. Different from the synthetic networks, the structure of diffusion subgraphs generated from different sources on a real network may be significantly different. This makes it much more difficult to infer the source node, especially for those algorithms which only consider one-dimension feature, e.g., DC, JC, RC and so on. While our proposed method, MCGNN, can capture the structural features from different levels, and can significantly improve the detection accuracy.

We first consider the case of small diffusion subgraphs. Figure 1.7a–c show the results on Email-univ networks. We see that MCGNN and GCN noticeably outperform other baseline methods on *Precision* and *Error Distance* (Fig. 1.7a, b). The similar results can also be found in Facebook and US Power Grid networks. It should be noticed that the phenomena observed here are quite complicated. First, for different real-world networks, sometimes the probabilistic likelihood estimation

Table 1.3 The performance of single source detection on the synthetic networks under three metrics, the bold values represent the best results

	Precision			Error distance			Normalized ranking		
	400	500	600	400	500	600	400	500	600
Scale-free									
DC	0.0	0.01	0.0	2.08	2.07	2.03	0.5536	0.5561	0.5445
JC	0.0	0.0	0.0	2.51	2.65	2.62	0.5318	0.4936	0.4487
RC	0.01	0.01	0.0	2.09	2.14	2.62	0.4852	0.4987	0.5106
DI	0.0	0.01	0.0	2.14	2.11	2.17	0.6283	0.6006	0.5885
RI	0.0	0.01	0.0	2.36	2.2	2.3	0.5592	0.5441	0.5272
DMP	0.39	0.25	0.18	1.22	1.84	1.87	0.2331	0.2693	0.3571
GSBA	0.0	0.0	0.0	2.84	2.86	2.9	0.6427	0.5974	0.6103
EPA	0.0	0.01	0.0	2.11	2.12	2.12	0.4949	0.5078	0.5149
GCN	0.41	0.33	0.29	1.02	1.57	1.78	0.1789	0.2219	0.2732
MCGNN	**0.45**	**0.37**	**0.31**	**0.89**	**1.23**	**1.60**	**0.1543**	**0.2071**	**0.2415**
ER random									
DC	0.16	0.13	0.02	1.31	1.62	1.75	0.0899	0.112	0.2317
JC	0.0	0.0	0.02	2.88	3.04	3.11	0.3134	0.3691	0.4355
RC	0.69	0.56	0.3	0.41	0.72	1.56	**0.0106**	0.0387	**0.0439**
DI	0.09	0.09	0.04	1.55	1.79	2.38	0.1093	0.1291	0.2462
RI	0.03	0.05	0.02	2.4	2.59	2.91	0.3236	0.4041	0.4441
DMP	0.62	0.4	0.35	0.73	1.06	1.02	0.1829	0.2141	0.203
GSBA	0.01	0.0	0.0	3.63	3.74	3.74	0.4888	0.5622	0.5628
EPA	0.5	0.37	0.11	0.81	1.2	2.04	0.0375	0.0631	0.1631
GCN	0.63	0.54	0.48	0.47	0.77	1.13	0.0128	0.0379	0.0512
MCGNN	**0.73**	**0.61**	**0.54**	**0.28**	**0.43**	**0.72**	0.0117	**0.0329**	0.0443
4-regular									
DC	0.31	0.23	0.2	1.09	1.4	1.75	0.0171	0.0198	0.0316
JC	0.07	0.06	0.04	2.74	3.57	4.23	0.074	0.118	0.158
RC	0.43	0.33	0.18	0.78	1.04	1.61	0.0107	0.0112	0.0238
DI	0.21	0.19	0.15	1.45	1.78	1.88	0.0214	0.0244	0.0288
RI	0.24	0.07	0.06	1.38	2.06	2.69	0.0244	0.0337	0.0801
DMP	0.42	0.43	**0.55**	0.72	0.75	**0.67**	0.01487	0.0199	0.0176
GSBA	0.0	0.0	0.0	6.01	5.9	5.88	0.5039	0.5258	0.4842
EPA	0.45	0.44	0.23	0.75	0.94	1.64	0.0097	0.0115	0.0188
GCN	0.51	0.47	0.41	0.59	0.71	0.87	0.0082	0.01375	0.01598
MCGNN	**0.67**	**0.53**	0.47	**0.43**	**0.62**	0.78	**0.0063**	**0.0097**	**0.0145**

And we repeat 100 independent experiments on the large diffusion subgraphs and report the average result

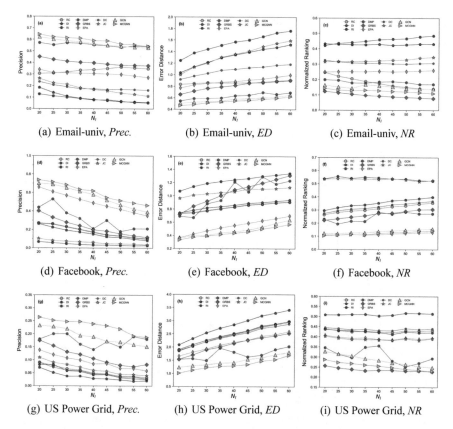

Fig. 1.7 The performance of different methods under all the three metrics on the real-world networks. We test all the algorithms 1000 times and report the results on average. Real-world networks include (**a**)–(**c**) Email-univ, (**d**)–(**f**) Facebook and (**g**)–(**i**) US Power Grid

based methods perform better (e.g., DMP, GSBA on Email-univ and US Power Grid), and sometimes the topological feature based methods perform better (e.g., EPA on Facebook). That means it is impossible to accurately infer the location of the source by following a single feature. Second, *Normalized Ranking* behaves very steadily as the number of infected nodes increases, as shown in Fig. 1.7c, f, i, and our proposed method can not perform the best in this case. Third, on Email-univ and Facebook, our method can achieve an average *Precision* of 70 %. However, on US Power Grid, although our method still performs the best, the overall average *Precision* is only 30% (as shown in Fig. 1.7g). These results indicate that the detection of the source in real networks is a difficult task, which requires more robust estimation algorithms in the future.

Similarly, we perform the experiments on large diffusion subgraphs with the same settings. Table 1.4 reports the results on the three real-world networks. Our method still can obtain the relative optimal results. However, we can see that the

Table 1.4 The performance of single source detection on the real-world networks under three metrics, the bold values represent the best results

	Precision			Error distance			Normalized ranking		
	400	500	600	400	500	600	400	500	600
Email-univ									
DC	0.0	0.0	0.02	2.07	2.17	2.26	0.5187	0.5504	0.5787
JC	0.02	0.0	0.0	2.46	2.75	2.74	0.477	0.4947	0.5096
RC	0.13	0.08	0.04	1.63	1.9	2.1	**0.1464**	**0.2149**	**0.2661**
DI	0.0	0.0	0.0	2.39	2.53	2.57	0.5784	0.5942	0.6141
RI	0.01	0.01	0.0	2.85	2.91	2.88	0.515	0.5507	0.468
DMP	0.3	0.27	**0.29**	1.49	1.61	1.68	0.4245	0.3807	0.3343
GSBA	0.0	0.01	0.0	3.02	3.09	3.06	0.662	0.6724	0.6271
EPA	0.08	0.12	0.04	1.86	1.92	2.15	0.4048	0.4624	0.4895
GCN	0.42	0.34	0.24	0.83	0.93	1.21	0.2138	0.2712	0.3126
MCGNN	**0.43**	**0.36**	0.27	**0.76**	**0.89**	**1.07**	0.1531	0.2382	0.2835
Facebook									
DC	0.0	0.0	0.0	1.19	1.33	1.29	0.5388	0.5066	0.51
JC	0.0	0.0	0.0	1.36	1.33	1.29	0.4399	0.4558	0.4491
RC	0.0	0.0	0.0	1.23	1.31	1.27	0.4901	0.4588	0.4841
DI	0.0	0.0	0.0	1.17	**1.18**	**1.19**	0.5257	0.4579	0.455
RI	0.0	0.0	0.0	1.43	1.42	1.36	0.4779	0.5215	0.4836
DMP	0.17	0.07	0.08	1.33	1.62	1.71	0.3102	0.3385	0.4137
GSBA	0.01	0.0	0.0	2.19	2.11	2.36	0.5764	0.5338	0.5338
EPA	0.0	0.0	0.0	1.11	1.25	1.23	0.3865	0.4011	0.4129
GCN	0.16	0.12	0.09	0.92	1.23	1.57	0.2891	0.3327	0.3573
MCGNN	**0.21**	**0.15**	**0.11**	**0.87**	1.19	1.32	**0.2132**	**0.2934**	**0.3427**
US Power Grid									
DC	0.0	0.0	0.0	5.19	5.53	5.48	0.48	0.4712	0.4359
JC	0.0	0.0	0.0	5.85	5.72	5.77	0.4373	0.3927	0.4016
RC	0.01	0.0	0.0	5.91	6.25	6.12	0.4728	0.5106	0.4471
DI	0.0	0.0	0.0	6.99	7.07	7.55	0.524	0.502	0.5019
RI	0.0	0.0	0.01	5.41	5.81	5.51	0.4521	0.4673	0.3851
DMP	0.10	0.06	**0.07**	3.42	4.57	4.98	0.3146	0.3568	0.3971
GSBA	0.0	0.01	0.0	9.16	9.28	10.29	0.4838	0.4453	0.4948
EPA	0	0.01	0	4.91	5.06	5.17	0.4371	0.4234	**0.3807**
GCN	0.11	0.07	0.02	3.71	4.23	4.68	0.3218	0.3568	0.3913
MCGNN	**0.15**	**0.1**	0.05	**3.02**	**3.84**	**4.19**	**0.3014**	**0.3316**	0.3893

We repeat 100 independent experiments on the large diffusion subgraphs and report the average result here

performance is lower than that on the synthetic networks, and other comparison methods can not effectively detect the source node. Therefore, it is still a big challenge to improve the detection precision for large-scale networks in the actual scenario.

1.6 Conclusion

In this chapter, we reviewed the problem of single information source detection on complex networks, and designed a multi-channel graph neural network framework to solve this problem. Experiments on both synthetic and real-world networks demonstrate the superiority of our method, which extracts features from different dimensions, i.e., node and edge channels.

In the future work, we may improve our method from the following three directions. First, we can extend the framework to multiple sources detection problem, to adapt more complex scenarios under unknown diffusion models. Second, in reality, we often lack the global information of the propagation path, thus we can exploit more robust GNN based algorithms to infer single or multiple sources under limited observation. Third, we can employ the real spreading data instead of the simulation data, to make our method potentially useful in practice.

References

1. Ali, S.S., Anwar, T., Rastogi, A., Rizvi, S.A.M.: EPA: Exoneration and prominence based age for infection source identification. In: Proceedings of the 28th ACM International Conference on Information and Knowledge Management, pp. 891–900 (2019)
2. Ali, S.S., Anwar, T., Rizvi, S.A.M.: A revisit to the infection source identification problem under classical graph centrality measures. Online Soc. Netw. Media **17**, 100061 (2020)
3. Altarelli, F., Braunstein, A., DallAsta, L., Ingrosso, A., Zecchina, R.: The patient-zero problem with noisy observations. J. Stat. Mech: Theory Exp. **2014**(10), P10016 (2014)
4. Altarelli, F., Braunstein, A., DallAsta, L., Lage-Castellanos, A., Zecchina, R.: Bayesian inference of epidemics on networks via belief propagation. Phys. Rev. Lett. **112**(11), 118701 (2014)
5. Barabási, A.L., Albert, R.: Emergence of scaling in random networks. Science **286**(5439), 509–512 (1999)
6. Batagelj, V., Zaversnik, M.: An o (m) algorithm for cores decomposition of networks. arXiv preprint cs/0310049 (2003)
7. Battaglia, P., Pascanu, R., Lai, M., Rezende, D.J., et al.: Interaction networks for learning about objects, relations and physics. In: Advances in Neural Information Processing Systems (2016), pp. 4502–4510
8. Bronstein, M.M., Bruna, J., LeCun, Y., Szlam, A., Vandergheynst, P.: Geometric deep learning: going beyond Euclidean data. IEEE Signal Process Mag. **34**(4), 18–42 (2017)
9. Carmi, S., Havlin, S., Kirkpatrick, S., Shavitt, Y., Shir, E.: A model of internet topology using k-shell decomposition. Proc. Natl. Acad. Sci. **104**(27), 11150–11154 (2007)
10. Chang, B., Chen, E., Zhu, F., Liu, Q., Xu, T., Wang, Z.: Maximum a posteriori estimation for information source detection. IEEE Trans. Syst. Man Cybern.: Syst. **50**(6), 2242–2256 (2018)

11. Choi, J., Moon, S., Woo, J., Son, K., Shin, J., Yi, Y.: Information source finding in networks: Querying with budgets. IEEE/ACM Trans. Networking **28**(5), 2271–2284 (2020)
12. Comin, C.H., da Fontoura Costa, L.: Identifying the starting point of a spreading process in complex networks. Phys. Rev. E **84**(5), 056105 (2011)
13. Dong, W., Zhang, W., Tan, C.W.: Rooting out the rumor culprit from suspects. In: Proceedings of the 2013 IEEE International Symposium on Information Theory, pp. 2671–2675. IEEE, New York (2013)
14. Dong, M., Zheng, B., Quoc Viet Hung, N., Su, H., Li, G.: Multiple rumor source detection with graph convolutional networks. In: Proceedings of the 28th ACM International Conference on Information and Knowledge Management, pp. 569–578 (2019)
15. ERDdS, P., R&wi, A.: On random graphs i. Publ. Math. Debrecen **6**(290–297), 18 (1959)
16. Fioriti, V., Chinnici, M., Palomo, J.: Predicting the sources of an outbreak with a spectral technique. Appl. Math. Sci. **8**(135), 6775–6782 (2014)
17. Fout, A., Byrd, J., Shariat, B., Ben-Hur, A.: Protein interface prediction using graph convolutional networks. In: Advances in Neural Information Processing Systems, pp. 6530–6539 (2017)
18. Goyal, P., Ferrara, E.: Graph embedding techniques, applications, and performance: A survey. Knowledge-Based Syst. **151**, 78–94 (2018)
19. Hamaguchi, T., Oiwa, H., Shimbo, M., Matsumoto, Y.: Knowledge transfer for out-of-knowledge-base entities: a graph neural network approach. In: Proceedings of the 26th International Joint Conference on Artificial Intelligence, pp. 1802–1808 (2017)
20. Hamilton, W., Ying, Z., Leskovec, J.: Inductive representation learning on large graphs. In: Advances in Neural Information Processing Systems, pp. 1024–1034 (2017)
21. Hirsch, J.E.: An index to quantify an individual's scientific research output. Proc. Natl. Acad. Sci. **102**(46), 16569–16572 (2005)
22. Jain, A., Borkar, V., Garg, D.: Fast rumor source identification via random walks. Social Network Anal. Mining **6**(1), 62 (2016)
23. Jiang, J., Wen, S., Yu, S., Xiang, Y., Zhou, W.: K-center: An approach on the multi-source identification of information diffusion. IEEE Trans. Inf. Forensics Secur. **10**(12), 2616–2626 (2015)
24. Jordan, C.: Sur les assemblages de lignes. J. Reine Angew. Math **70**(185), 81 (1869)
25. Kermack, W.O., McKendrick, A.G.: Contributions to the mathematical theory of epidemics II. the problem of endemicity. Proc. R. Soc. London Ser. A, Containing Papers of a Mathematical and Physical Character **138**(834), 55–83 (1932)
26. Khalil, E., Dai, H., Zhang, Y., Dilkina, B., Song, L.: Learning combinatorial optimization algorithms over graphs. In: Advances in Neural Information Processing Systems, pp. 6348–6358 (2017)
27. Kipf, T.N., Welling, M.: Semi-supervised classification with graph convolutional networks. arXiv preprint arXiv:1609.02907 (2016)
28. Leskovec, J., Mcauley, J.J.: Learning to discover social circles in ego networks. In: Advances in Neural Information Processing Systems, pp. 539–547 (2012)
29. Lokhov, A.Y., Mézard, M., Ohta, H., Zdeborová, L.: Inferring the origin of an epidemic with a dynamic message-passing algorithm. Phys. Rev. E **90**(1), 012801 (2014)
30. Luo, W., Tay, W.P.: Finding an infection source under the sis model. In: Proceedings of the 2013 IEEE International Conference on Acoustics, Speech and Signal Processing, pp. 2930–2934. IEEE, New York (2013)
31. Luo, W., Tay, W.P., Leng, M.: Identifying infection sources and regions in large networks. IEEE Trans. Signal Process. **61**(11), 2850–2865 (2013)
32. Luo, W., Tay, W.P., Leng, M.: How to identify an infection source with limited observations. IEEE J. Sel. Top. Signal Process. **8**(4), 586–597 (2014)
33. Page, L., Brin, S., Motwani, R., Winograd, T.: The pagerank citation ranking: bringing order to the web. Technical report, Stanford InfoLab (1999)
34. Park, N., Kan, A., Dong, X.L., Zhao, T., Faloutsos, C.: Multiimport: inferring node importance in a knowledge graph from multiple input signals. In: Proceedings of the 26th ACM SIGKDD International Conference on Knowledge Discovery and Data Mining, pp. 503–512 (2020)

35. Pastor-Satorras, R., Castellano, C., Van Mieghem, P., Vespignani, A.: Epidemic processes in complex networks. Rev. Mod. Phys. **87**(3), 925 (2015)
36. Prakash, B.A., Vreeken, J., Faloutsos, C.: Spotting culprits in epidemics: how many and which ones? In: Proceedings of the 2012 IEEE 12th International Conference on Data Mining, pp. 11–20. IEEE, New York (2012)
37. Rossi, R.A., Ahmed, N.K.: The network data repository with interactive graph analytics and visualization. In: AAAI (2015). http://networkrepository.com
38. Ruan, Z., Iniguez, G., Karsai, M., Kertész, J.: Kinetics of social contagion. Phys. Rev. Lett. **115**(21), 218702 (2015)
39. Ruan, Z., Tang, M., Gu, C., Xu, J.: Epidemic spreading between two coupled subpopulations with inner structures. Chaos: An Interdiscip. J. Nonlinear Sci. **27**(10), 103104 (2017)
40. Ruan, Z., Yu, B., Shu, X., Zhang, Q., Xuan, Q.: The impact of malicious nodes on the spreading of false information. Chaos: An Interdiscip. J. Nonlinear Sci. **30**(8), 083101 (2020)
41. Sanchez-Gonzalez, A., Heess, N., Springenberg, J.T., Merel, J., Riedmiller, M.A., Hadsell, R., Battaglia, P.: Graph networks as learnable physics engines for inference and control. In: ICML (2018)
42. Scarselli, F., Gori, M., Tsoi, A.C., Hagenbuchner, M., Monfardini, G.: The graph neural network model. IEEE Trans. Neural Networks **20**(1), 61–80 (2008)
43. Seidman, S.B.: Network structure and minimum degree. Social Netw. **5**(3), 269–287 (1983)
44. Shah, D., Zaman, T.: Detecting sources of computer viruses in networks: theory and experiment. In: Proceedings of the ACM SIGMETRICS International Conference on Measurement and Modeling of Computer Systems, pp. 203–214 (2010)
45. Shah, D., Zaman, T.: Rumors in a network: who's the culprit? IEEE Trans. Inf. Theory **57**(8), 5163–5181 (2011)
46. Shah, C., Dehmamy, N., Perra, N., Chinazzi, M., Barabási, A.L., Vespignani, A., Yu, R.: Finding patient zero: learning contagion source with graph neural networks. arXiv preprint arXiv:2006.11913 (2020)
47. Veličković, P., Cucurull, G., Casanova, A., Romero, A., Lio, P., Bengio, Y.: Graph attention networks. arXiv preprint arXiv:1710.10903 (2017)
48. Wang, Z., Dong, W., Zhang, W., Tan, C.W.: Rooting our rumor sources in online social networks: the value of diversity from multiple observations. IEEE J. Sel. Top. Signal Process. **9**(4), 663–677 (2015)
49. Wang, Z., Wang, C., Pei, J., Ye, X.: Multiple source detection without knowing the underlying propagation model. In: Proceedings of the Thirty-First AAAI Conference on Artificial Intelligence (2017)
50. Watts, D.J., Strogatz, S.H.: Collective dynamics of small-world networks. Nature **393**(6684), 440–442 (1998)
51. Wu, Z., Pan, S., Chen, F., Long, G., Zhang, C., Philip, S.Y.: A comprehensive survey on graph neural networks. IEEE Trans. Neural Networks Learn. Syst. **32**(1), 4–24 (2021)
52. Zhang, Z.K., Liu, C., Zhan, X.X., Lu, X., Zhang, C.X., Zhang, Y.C.: Dynamics of information diffusion and its applications on complex networks. Phys. Rep. **651**, 1–34 (2016)
53. Zhang, Z., Cui, P., Zhu, W.: Deep learning on graphs: a survey. IEEE Trans. Knowl. Data Eng. (2020)
54. Zhou, D., Bousquet, O., Lal, T.N., Weston, J., Schölkopf, B.: Learning with local and global consistency. In: Advances in Neural Information Processing Systems, pp. 321–328 (2004)
55. Zhu, K., Ying, L.: Information Theory and Applications Workshop (ITA), San diego, CA (2013)
56. Zhu, K., Ying, L.: Information source detection in the sir model: a sample-path-based approach. IEEE/ACM Trans. Netw. **24**(1), 408–421 (2014)
57. Zhu, K., Chen, Z., Ying, L.: Catch'em all: Locating multiple diffusion sources in networks with partial observations. In: Proceedings of the 31st AAAI Conference on Artificial Intelligence (AAAI 2017), pp. 1676–1682. AAAI press, New York (2017)

Chapter 2
Link Prediction Based on Hyper-Substructure Network

Jian Zhang, Jinyin Chen, and Qi Xuan

Abstract Link prediction has long been the focus in the analysis of network-structured data. Though straightforward and efficient, heuristic approaches like *Common Neighbors* perform link prediction with pre-defined assumptions and only use superficial structural features. While it is widely acknowledged that a node could be characterized by a bunch of neighbor nodes, network embedding algorithms and newly emerged graph neural networks still exploit structural features on the whole network, which may inevitably bring in noises and limits the scalability of those methods. In this chapter, we propose an end-to-end deep learning framework, namely *hyper-substructure enhanced link predictor* (HELP), for link prediction. HELP utilizes local topological structures from the neighborhood of the given node pairs, avoiding useless features. For further exploiting higher-order structural information, HELP also learns features from *hyper-substructure network* (HSN). Extensive experiments on five benchmark datasets have shown the state-of-the-art performance of HELP on link prediction.

2.1 Introduction

As a representative network analysis task, link prediction infers the linkage status of a given pair of nodes in a network. Due to its practicability, link prediction has been widely applied in various areas, such as commodity recommendation on e-commerce platforms and friends recommendation in online social networks. In the study of biological networks, such as protein-protein interaction (PPI) network, it would be costly to find an underlying interaction between proteins through laboratorial experiments since there are countless candidates. Link prediction offers

J. Zhang · J. Chen · Q. Xuan (✉)
Institute of Cyberspace Security, Zhejiang University of Technology, Hangzhou, China

College of Information Engineering, Zhejiang University of Technology, Hangzhou, China
e-mail: xuanqi@zjut.edu.cn

Q. Xuan et al. (eds.), *Graph Data Mining*, Big Data Management,
https://doi.org/10.1007/978-981-16-2609-8_2

a list of the interactions that most likely exist, which could effectively reduce economic and time costs.

There have been a bunch of works that focus on link prediction. Heuristic link prediction approaches, such as *Common Neighbors* (CN), *Resource Allocation Index* (RA) and Katz Index, use either local or global similarity scores to make prediction [17]. For instance, in OSN like Weibo, candidate friends would be recommended according to the common friends one has with other users. Though straightforward and efficient, the pre-defined assumption of heuristic methods maybe fail sometimes. For instance, researcher assumes that the users with more common friends are more likely to form friendship in social networks. However, the assumption fails when it comes to PPI networks. It has been proved that proteins sharing more common neighbors have less interactions [15]. Heuristic approaches only learn superficial structural features but fail to characterize the complexity of networks. Later developed network embedding algorithms, such as DeepWalk [21] and node2vec [7], focus on the contexts of nodes in the sequences generated by random walk or its variants. The random walk-based algorithms, however, need to learn embeddings over the whole network, which limits the scalability to large-scale networks even with parallelized computation. Apart from the need for plenty of pre-defined parameters, they are gradually exceeded by newly emerged graph neural networks (GNNs).

With *graph convolution network* (GCN) [14] as the representative, the newly developed GNNs have shown their power on many tasks. Despite the performance, most GNNs make inference over the whole network, which leads to high computational complexity and might bring in noise since not all nodes are useful for downstream tasks. Though graph attention network (GAT) [27] and GraphSAGE [8] try to learn embeddings over the neighborhood of nodes, they still focus on superficial features. And so does SEAL [32]. In social networks, a few individuals may consists of a group for some purposes. And similar cases like pairwise programming are also common in open source software development. The interactions between the groups could characterize the higher-order structural features of the network. However, few existing methods pay attention to such information.

To tackle the limitations of existing algorithms, we propose an end-to-end deep learning framework, namely *hyper-substructure enhanced link predictor* (HELP), for link prediction. And part of the work has been published in [34]. Converting link prediction problem to graph classification, HELP infers linkage status of a given pair of nodes based on their neighborhood rather than the whole network. The neighborhood selected by personalized PageRank (PPR) [3] consists of the nodes closest to the given nodes, avoiding useless structural features. Neighborhood learning enables HELP with scalability to large-scale networks. And node feature vectors characterizing the relative position of the node in the network are constructed to improve the accuracy. To utilize higher-order topological structures, networks constructed by substructures of the subgraph, namely *hyper-substructure network* (HSN), are also used for link prediction. And we use second-order HSN to validate its advantage in this paper. The obtained networks are finally fed into a

GNN to make prediction. The main contributions of the paper are summarized as follows:

- We propose an end-to-end deep learning framework, namely hyper-substructure enhanced link predictor (HELP), for link prediction. And its outstanding performance has been proved by extensive experiments.
- We convert link prediction problem to graph classification and infer linkage status based on subgraphs rather than the whole network, expending the scalability of HELP.
- We creatively introduce HSN into link prediction to utilize higher-order structural information, which offers insights of network structure mining.

2.2 Existing Link Prediction Methods

2.2.1 Heuristic Methods

Heuristic link prediction approaches define similarity indices representing the likelihood that a link exist between a node pair. The indices could be local or global. Local similarity indices use local structures, such as first-order neighbors, to measure the similarity of nodes. Among all the heuristic approaches, Common Neighbors (CN) [19] is the simplest. CN assumes that the target node pair u and v are more likely to form a link if they have more common neighbors which is defined as

$$s_{u,v}^{CN} = |\Gamma(u) \cap \Gamma(v)|, \tag{2.1}$$

where $\Gamma(u)$ and $\Gamma(v)$ represent the first-order neighbors of u and v, respectively. In social networks, the amount of common neighbors between the nodes with higher degrees may be hundreds times of it between the nodes of small degrees. However, we could not justify that the probability of link between the nodes of higher degrees is higher due to the large mount of neighbors they have. To overcome this shortage, Jaccard Index [9] proposes to normalize CN by the total amount of neighbors, which is defined as

$$s_{u,v}^{Jaccard} = \frac{|\Gamma(u) \cap \Gamma(v)|}{|\Gamma(u) \cup \Gamma(v)|}. \tag{2.2}$$

From another perspective, Salton Index [23] divides CN by the square root of the product of the degrees of u and v. Its definition goes as

$$s_{u,v}^{Salton} = \frac{|\Gamma(u) \cap \Gamma(v)|}{\sqrt{k_u \times k_v}}, \tag{2.3}$$

where k_u denotes the degree of u. Both Jaccard Index and Salton Index are developed form CN. And many other approaches, such as Hub Promoted Index (HPI) [22] and Hub Depressed Index (HDI), are also associated with CN but with different normalization methods. Motivated by the resource allocation dynamics on networks, Resource Allocation Index (RA) [35] regards the neighbors as transmitters which send some resource from u to v. The definition is given by

$$s_{u,v}^{RA} = \sum_{z \in \Gamma(u) \cap \Gamma v} \frac{1}{k_z}. \tag{2.4}$$

Local similarity indices characterize the closeness between nodes by local structures, while most global similarity indices are associated with network paths. Katz Index [11] ensembles all paths between u and v. The mathematical expression of Katz Index reads

$$s_{u,v}^{katz} = \sum_{l=1}^{\inf} \beta^l |paths_{u,v}^{(l)}| = \beta A_{u,v} + \beta^2 A_{u,v}^2 + \cdots, \tag{2.5}$$

where $paths_{u,v}^{(l)}$ represents all the paths between u and v of length l and β is the damping weight of paths of different lengths. And A denotes the adjacent matrix of the network. There are many other path-based similarity indices, such as Leicht-Holme-Newman Index (LHN2) [16] and SimRank [10], whose effectiveness has been proved under some particular scenarios. PageRank [4], as a representative of path-based method, has been widely applied in online web search engine. And it is defined as

$$PR(u) = c \sum_{v \in B_u} \frac{PR}{L(v)} + \frac{1-c}{N}, \tag{2.6}$$

where B_u represents the nodes that points to u, $L(v)$ denotes the number of nodes that v points to and N is the number of nodes. $PR(u)$ defines the influence of u and also gives the probability whether u will form links with other nodes. Though heuristic link prediction approaches are simple and efficient, they may fail for some networks since they cannot fully utilize the complex network structure.

2.2.2 Embedding-Based Methods

To learn complex network structural features, DeepWalk [21] creatively introduces Word2Vec [18] into network representation that mapping random walk sequences to vectors. It first samples a certain amount of node sequences by applying random walk starting at different nodes and then uses skip-gram to embed each node into vector space. Node2vec [7] extends DeepWalk by incorporating depth-first

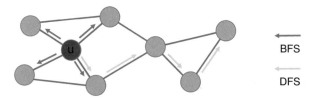

Fig. 2.1 The walking strategies for node2vec

Table 2.1 Binary operators for feature transformation

Operator	Symbol	Definition
Average	$\boxplus$	$[f(u) \boxplus f(v)]_i = \frac{f_i(u) + f_i(v)}{2}$
Hadamard	$\boxdot$	$[f(u) \boxdot f(v)]_i = f_i(u) * f_i(v)$
Weighted-L1	$\|\cdot\|_{\bar{1}}$	$\|f_i(u) \cdot f_i(v)\|_{\bar{1}i} = \|f_i(u) - f_i(v)\|$
Weighted-L2	$\|\cdot\|_{\bar{2}}$	$\|f_i(u) \cdot f_i(v)\|_{\bar{2}i} = \|f_i(u) - f_i(v)\|^2$

search (DFS) and breadth-first search (BFS) to learn both local and global structural features. As is shown in Fig. 2.1, the walk starts at u and the preference of DFS or BFS is controlled by two parameters, p and q. Apparently, node2vec will degenerate to DeepWalk when $p = q = 1$.

DeepWalk and node2vec are two typical examples developed from random walk. Later developed random walk-based methods, such as metapath2vec [6] and netwalk [31], are designed for heterogeneous networks or dynamic networks. The approaches above map nodes from non-Euclidean space into vector space. To perform link prediction, the node embeddings should be transformed into link representations. Table 2.1 presents four common operators to transfer node embeddings to link vectors. Researchers could also define other operators, such as concatenation, to meet their own needs. After obtaining link vectors, machine learning algorithms like logistic regression and SVM could be applied to solve the link prediction problem.

Instead of describing network structure in node sequences generated by random walk, LINE [26] jointly optimizes the first-order and second-order proximity to learn network embeddings. And the proximity is defined as,

$$O_1 = d(\hat{p}_1(\cdot, \cdot), p_1(\cdot, \cdot)),$$
$$O_2 = \sum_{v_i \in V} \lambda_i d(\hat{p}_1(\cdot, c\cdot), p_1(\cdot, c\cdot)), \tag{2.7}$$

where O_1 and O_2 represent the first-order and second-order proximity, respectively. In Eq. (2.7), $p_1(\cdot, \cdot)$ denotes the joint distribution of the latent embeddings while $\hat{p}_1(\cdot, \cdot)$ is the empirical distribution. And $d(\cdot, \cdot)$ is the distance between two distributions.

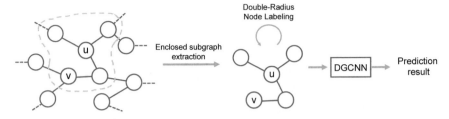

Fig. 2.2 The overall framework of SEAL

2.2.3 Deep Learning-Based Models

To automatically capture structural features, a bunch of graph neural networks are proposed. Earlier approaches mainly employ multi-layer perceptrons to build GAEs for network embedding learning. Deep Neural Network for Graph Representations (DNGR) [5] uses a stacked denoising autoencoder [28] to encode and decode the PPMI matrix via multi-layer perceptrons. Concurrently, Structural Deep Network Embedding (SDNE) [29] uses a stacked autoencoder to preserve the node first-order proximity and second-order proximity jointly. Rather than processing networks in spatial domain, Kipf et al. [14] analyze networks in Fourier domain and propose graph convolutional network (GCN) to solve the node classification problem. An extension of GCN, graph auto-encoder (GAE) [13], makes predictions in an end-to-end way. From spatial perspective, GCN aggregates several specific nodes to represent a target node. GraphSage [8] and graph attention network [27] which introduce attention mechanism into network representation learning also follow the idea but address the problem in different manners. Learning over the whole network limits the scalability of algorithms. Diffusion Convolutional Neural Network (DCNN) [1] regards graph convolutions as a diffusion process. It assumes information is transferred from one node to one of its neighboring nodes with a certain transition probability so that information distribution can reach equilibrium after sevenral rounds. To overcome the existing drawback, SEAL [32] first extracts an enclosed subgraph for the target node pair and then makes prediction based on the subgraph. To characterize the relative positions and importance with the subgraph, Double-Radius Node Labeling (DRNL) is proposed to assign each node with a label of which the one-hot encoding vector is regarded as the feature of the corresponding node. The final link prediction result is given by a graph classification called DGCNN [33] (Fig. 2.2).

2.3 Methodology

Heuristic approaches as well as random walk-based algorithms have suggested that node structural features could be characterized by its neighborhood. Motivated by this potential, we introduce an end-to-end link prediction model based on hyper-substructure network of which the effectiveness will be proved in Sect. 2.4. In this section, we first give a detailed description of the problem and then introduce the HELP model in three parts: neighborhood normalization, HSN construction and the graph neural network used for link prediction.

2.3.1 Problem Formulation

Suppose we have a graph $G = \langle V, E \rangle$ where $V = \{v_i | i = 0, 1, \cdots, N - 1\}$ denotes the set of N nodes and $E \subseteq V \times V$ represents the edge set. Given a pair of nodes u and v, our goal is to infer the linkage status between u and v based on their neighborhood $\Gamma(u, v)$. These neighbors of u and v consist of a subgraph centered at (u, v). And the HSN of (u, v) is constructed subsequently. Then we transfer the link prediction problem into graph classification problem.

2.3.2 Neighborhood Normalization

Before performing link prediction, we need to extract a subgraph of G for the target node pair (u, v). It is an intuitive idea to use the first-order or second-order neighbors to represent the node pair, since these nodes are close to the target nodes from the perspective of topology. But the size of such kind of subgraph varies when it comes to different nodes, making it difficult for HSN construction. And not all first-order or second-order neighbors could contribute to link prediction. Rather than directly using the topologically close neighbors of u and v, we propose to use personalized PageRank (PPR) to extract *key* neighbors for (u, v).

As powerful as GNNs are, researchers may find it difficult to apply them to large-scale networks due to the high computation complexity. Bojchevski et al. [3] propose to employ PPR to approximate GCN and significantly improve the efficiency of the model without loss of accuracy. In the model, PPR assigns each node pair with a probability that one node would travel to the other. The higher the probability is, the closer the two nodes are. Likewise, we apply PPR to obtain the neighbors that closest to the target nodes. The definition of PPR goes as

$$\Pi^{ppr} = \alpha(I_n - (1 - \alpha)D^{-1}A)^{-1}, \tag{2.8}$$

where A is the adjacent matrix of G, D is the degree matrix and α represents the restart probability. Each row $\pi(i) = \Pi^{ppr}_{(i)}$ is the PPR vector for node i and the element $\pi(i)_j$ reflects the closeness between node i and node j. We then sort $\pi(i)$ in descending order of the element value and choose the first N_{nb} nodes to construct the subgraph which is denoted by $G(u, v)$. For the nodes with same PPR values, we make random selections among them, ensuring that the number of neighbors reaches N_{nb}. Following the steps, we obtain $\Gamma(u)$ and $\Gamma(v)$ for nodes u and v, respectively. Figure 2.3 gives an example of the selection of $\Gamma(u)$ in which $N_{nb} = 4$.

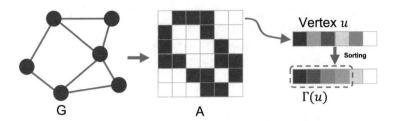

Fig. 2.3 An example of selecting neighbors according to PPR

2.3.3 HSN Construction

After obtaining $\Gamma(u)$ and $\Gamma(v)$, we then need to transform them into one unified subgraph to characterize the structural feature of the target node pair (u, v). Firstly, we link the nodes in $\Gamma(u)$ if they are originally linked in G, which constructs G_u. And we could get G_v following the same operation. G_u and G_v characterize the structural features of nodes u and v, respectively. To present the closeness of nodes u and v, we also link the nodes in G_u and G_v if they are connected in G. If G_u and G_v contain the same nodes, we consider them as different nodes and then connect them in $G(u, v)$. The operation leads to virtual nodes and virtual links in $G(u, v)$ but keeps the size of $G(u, v)$ always the same. And the more virtual links, the stronger $\Gamma(u)$ and $\Gamma(v)$ are connected, implying that u and v are more likely to form a link. Also, we construct a feature matrix for $G(u, v)$. For node i in $G(u, v)$, the feature vector is defined as the concatenation of one-hot encoded $LSP(i, u)$ and $LSP(i, v)$ on $G(u, v)$, where $LSP(i, u)$ denotes the length of shortest path between i and u on $G(u, v)$. Such definition of node feature describes the relative position between neighbors and (u, v), which could improve the accuracy of HELP.

$G(u, v)$ characterizes the local structural features of (u, v) but do not describes the interactions between small groups in the network. To exploit higher-order structural information, we propose HSN to characterize the interactions of the substructures in $G(u, v)$. Kth order HSN, denoted by $HSN^{(K)}$, is constructed by following steps:

1. **Node Grouping**. Group the nodes in $G(u)$ and make sure that each node is grouped with arbitrary node at least once. The groups are represented by $p^{(u)} = \{p_1^{(u)}, \cdots, p_M^{(u)}\}$ where $M = C_{N_{nb}}^K$ and $p_i^{(u)}$ consists of K nodes randomly selected from $G(u)$. Also, we can obtain $p^{(v)}$ following the same procedure. The nodes in each group are connected if they are connected in $G(u, v)$.
2. **Node Group Sorting**. Sort $p^{(u)}$ in ascending order of distance d which is defined as

$$d\left(p_i^{(u)}|G(u, v)\right) = \sum_{j=0}^{K} LSP\left(p_i^{(u)}(j), v\right). \tag{2.9}$$

The definition of d is the sum of LSP between the nodes in $p_i^{(u)}$ and v in $G(u, v)$. And we process $p^{(v)}$ in the same way but with

$$d\left(p_i^{(v)}|G(u, v)\right) = \sum_{j=0}^{K} LSP\left(p_i^{(v)}(j), u\right). \tag{2.10}$$

3. **Node Group Wiring**. We consider each group as a node of HSN. After sorting, we select the first N_H groups from $p^{(u)}$ and $p^{(v)}$, respectively. Given $p_i^{(u)}$ and $p_j^{(u)}$, there exists an edge between them if $\left|p_i^{(u)} \cap p_j^{(u)}\right| < \left|p_i^{(u)}\right| + \left|p_j^{(u)}\right|$; Given $p_i^{(u)}$ and $p_i^{(v)}$, they are connected if there is at least one pair of nodes $\left(p_i^{(u)}(a), p_i^{(v)}(b)\right)$ is connected on $G(u, v)$ (a and b are arbitrary node indices in corresponding group).

Following the above steps, we could obtain $HSN^{(K)}$. And we could also construct node features for the nodes in $HSN^{(K)}$ just as the way of the node feature construction in $G(u, v)$. Figure 2.4 gives an example of $HSN_{u,v}^{(2)}$ construction with $N_{nb} = 5$, $K = 2$ and $N_H = 2$. The groups with small d, such as (4, 5) and (6, 7), are used for the construction of $HSN^{(2)}$, which enhances the interactions between $G(u)$ and $G(v)$.

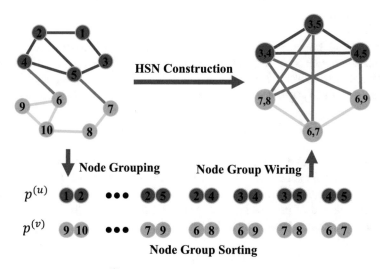

Fig. 2.4 Illustration of $HSN_{u,v}^{(2)}$'s construction. The red nodes represent the nodes in $G(u)$ and those of yellow are the nodes in $G(v)$. The blue lines denote the interactions between $G(u)$ and $G(v)$

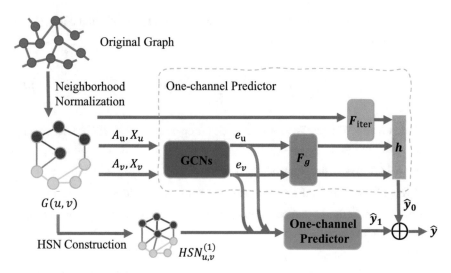

Fig. 2.5 The framework of HELP

2.3.4 HELP

After obtaining $G(u, v)$ and $HSN^{(K)}$ for (u, v), we then use an end-to-end deep learning framework, namely HELP, to infer whether there exists an edge between u and v or not.

The overall framework is given in Fig. 2.5. HELP consists of multiple *one-channel predictors* and each of them processes one HSN ($G(u, v)$ could be considered as $HSN^{(0)}$). We incorporate graph convolution network (GCN) [14] into *one-channel predictor* to learn node embeddings. One layer GCN is defined as

$$GCN(A, X) = \sigma(\tilde{D}^{-\frac{1}{2}} \tilde{A} \tilde{D}^{-\frac{1}{2}} X W), \tag{2.11}$$

where $\tilde{A} = A + I$ and $\tilde{D}$ is the normalized degree matrix. W is the weight matrix and σ refers to the activation function. Here we use $\sigma \equiv ReLU(\cdot) = max(0, \cdot)$. *One-channel predictor* processes $G(u)$, $G(v)$ and the interactions between them separately, instead of processing $G(u, v)$ in its entirety. With the incorporation of GCN, the *one-channel predictor* is described in Eq. (2.12):

$$e_u = GCNs(\tilde{A}_u, X_u),$$

$$e_v = GCNs(\tilde{A}_v, X_v),$$

$$o^u = FLATTEN(e_u),$$

$$o^v = FLATTEN(e_v),$$

$$h_u = F_g(e_u), \tag{2.12}$$

$$h_v = F_g(e_v),$$

$$h_{u \times v} = F_{iter}(A_{u \times v})$$

$$h = CONCAT(h_u, h_v, h_{u \times v})$$

$$\hat{y}_0 = softmax(W_{out}h + b),$$

where $\tilde{A}_u$ and X_u represent the normalized adjacency matrix and the feature matrix of $G(u)$, respectively. *One-channel predictor* first embeds the nodes in $G(u)$ and $G(v)$ with a two-layer GCN separately and then encode the embeddings with a multi-layer perception F_g. As for e_u and e_v, they represent the node embeddings of $G(u)$ and $G(v)$, respectively. And o^u as the rows' concatenation of e_u denotes the embedding vector of $G(u)$ and so does o^v. The interaction between $G(u)$ and $G(v)$, denoted by $A_{u \times v}$, is also encoded by F_{iter}. The prediction result $\hat{y}_0$ is given by a softmax classifier. To integrate higher-order structural features, we also make prediction on $HSN_{u,v}^{(2)}$ with *one-channel predictor* and obtain the corresponding prediction result $\hat{y}_1$. The final result $\hat{y}$ is given by

$$\hat{y} = \frac{1}{2}(\hat{y}_0 + \hat{y}_1). \tag{2.13}$$

It's worth noticing that we use the concatenation of e_u and e_v generated based on $G(u, v)$ as the features when we make inference based on $HSN_{u,v}^{(2)}$. And the framework can be easily extended with higher-order HSN.

The model then is optimized with Adam [12]. And the objective function L_{total} mainly contains two parts: cross-entropy error L_c and embedding similarity error L_s. The minimization of L_c directly ensures the link prediction performance of the model. And we also minimize KL-divergence to measure the similarity of o^u and o^v if there exists an edge between u and v, which could shorten the distance of the features of connected nodes. Consisting of the two parts, L_{total} is define as

$$L_c = -y \log \hat{y} + (1 - y) \log(1 - \hat{y})$$

$$L_s = \frac{1}{d} \sum_{i=0}^{d-1} y_i (o_i^u \log(o_i^u) - o_i^u \log(o_i^v)) \tag{2.14}$$

$$L_{total} = \gamma L_c + (1 - \gamma)L_s + \beta L_{reg},$$

where γ is the coefficient used for balancing L_c and L_s. L_{reg} represents L_2 regularization term and β is the weight decay coefficient.

2.4 Experiment

2.4.1 Datasets

We compare different link prediction algorithms on 8 benchmark datasets:

- **C.elegans** [30] is an network which models the neural interactions of the nematode worm C.elegans. The nodes represent the neurons and the links are the metabolic reactions. The original network is directed and weighted but we simplify it to an undirected and unweighted one in the experiments.
- **USAir** [2] is a network of US air transportation in which the nodes denote the airports and the links represent flights between different airports. Also, we transfer the original network to an undirected and unweighted one.
- **HP** is an undirected network which models the interactions between human proteins.
- **NetScience** [20] contains the collaborations between the scientists in network science area. The nodes are researchers and the links represent the co-authorship.
- **Power** [30] is an undirected network of power grid network where a vertex either represents a generator, a transformer or a substation and an edge is a power supply line.
- **Router** [25] is a network of router-level Internet in which the nodes are routers and the links represent the data transmission between routers.
- **Cora** and **Citeseer** [24] are networks modeling citation relationships between scientific papers.

In this chapter, we focus on the networks without direction and weights and the basic statistics are summarized in Table 2.2.

2.4.2 Link Prediction Methods for Comparison

We compare the proposed method with five baselines, including heuristic methods, random walk-based methods and deep learning-based methods.

- **Heuristic methods**: **RA**, **Jaccard Index** and **PR**. The definitions of the three similarity indices have been introduced in Sect. 2.2.

Table 2.2 Basic statistics of the datasets

	C.elegans	USAir	HP	NetScience	Power	Router	Citeseer	Cora
N	297	332	1706	1461	4941	5022	3279	2708
$|E|$	2148	2126	3191	2754	6594	6258	4552	5278
$\langle k \rangle$	14.46	12.81	3.72	3.75	2.67	2.49	2.78	3.90

$\langle k \rangle$ represents the average degree of the network

- **Embedding-based method**: **node2vec** and **LINE**. We first use node2vec and LINE to capture node structural features and then generate link features through Hardmard product. Finally, a logistic regression model is applied to make prediction.
- **Deep learning-based method**: **SEAL**. To ensure a fair comparison, we only use the subgraphs to make prediction when applying SEAL.

2.4.3 Evaluation Metrics

We employ Area Under the Curve (AUC) and Average Precision (AP) to evaluate the performance of different link prediction methods. And the definition of the two metrics are listed as follows.

- **AUC**: AUC could be regarded as the probability that a higher score would be assigned to a randomly chosen node pair if they form a link in the original network. And its definition is given as

$$AUC = \frac{n' + 0.5n''}{n}, \tag{2.15}$$

where n is the number of times that node pairs belonging to existing links get higher scores and n'' the number of times that the node pairs that do not form links. And n is total number of comparison times.
- **AP**: AP takes the average of the precision at each threshold, with its definition goes as

$$AP = \sum_{t}(R_t - R_{t-1})P_t, \tag{2.16}$$

where P_t and R_t are the precision and recall at threshold t, respectively.

2.4.4 Experimental Settings

In the experiments, each dataset is split into training and testing sets at the ratio of 4:1. Specifically, we choose 80% links as positive samples of training set and randomly sample the same number of nonexistent edges as negative training samples. The remaining 20% links are positive testing data and the same amount of nonexistent links are sampled as negative testing data. And the positive testing links are removed from the original network.

As for the baselines, heuristic methods need no training and pre-defined parameters. Thus we could make predictions directly according to the similarity indices. For node2vec, we fix the embedding dimension as 128 and the optimal key parameters p and q are obtained through grid search over $\{0.50, 0.75, 1.00, 1.25, 1.50\}$. After obtaining the embedding vectors of each node, we use Hardamard product to generate link features for each node pair. And then, an LR model is applied to make prediction. SEAL uses *auto* subgraph selection which means that 1-hop or 2-hop subgraphs will be automatically selected to achieve better performance. For the proposed model, HELP, we set $N_{nb} = N_H = 35$ for the construction of subgraph and $HSN^{(2)}$.

2.4.5   Link Prediction Results

All the experiments are conducted 10 times and the average performance is recorded to avoid contingency. As reported in Tables 2.3 and 2.4, HELP outperforms other baselines in most cases considering both *AUC* and *AP*, which shows the effectiveness and practicability of HELP. As expected, heuristic methods based on local structural similarities are not as good as other methods, especially on sparser networks like NS, HP and Router. It is because that first-order and second-

Table 2.3 Performance of different link prediction methods with regard to AUC (The best results are in bold.)

Dataset	RA	Jaccard	PR	node2vec	LINE	SEAL	HELP
C.elegans	0.8484	0.7761	0.8517	0.8547	0.7920	0.8371	**0.8553**
USAir	0.9405	0.8923	0.9169	0.9094	0.8352	0.9473	**0.9482**
HP	0.5190	0.5193	0.6567	0.6446	0.7182	0.8850	**0.8952**
NS	0.9096	0.9110	0.9221	0.8776	0.9717	0.9844	**0.9912**
Power	0.5750	0.5749	0.5951	0.8007	0.6796	0.8603	**0.8838**
Router	0.5537	0.5528	0.6479	0.5705	0.8246	0.9121	**0.9423**
Cora	0.6993	0.7014	0.8049	0.9091	0.8098	0.9181	**0.9291**
Citeseer	0.6539	0.6530	0.7323	0.8797	0.8269	0.8754	**0.8802**

Table 2.4 Performance of different link prediction methods with regard to AP (The best results are in bold.)

Dataset	RA	Jaccard	PR	node2vec	LINE	SEAL	HELP
C.elegans	0.8437	0.7321	0.8453	0.8399	0.7602	0.8371	**0.8445**
USAir	**0.9468**	0.8670	0.9345	0.9000	0.8141	0.9450	0.9364
HP	0.5204	0.5128	0.7284	0.6501	0.7396	0.8962	**0.9060**
NS	0.9097	0.9109	0.9261	0.9286	0.9780	0.9851	**0.9927**
Power	0.5745	0.5744	0.7377	0.8453	0.7222	0.8603	**0.8855**
Router	0.5536	0.5465	0.7577	0.6114	0.8497	0.8956	**0.9416**
Cora	0.7005	0.6978	0.8638	0.9274	0.8452	0.9298	**0.9404**
Citeseer	0.6539	0.6525	0.7858	0.8842	0.8657	0.8874	**0.9093**

order neighbors could not characterize the structural features of the target node. However, the local structure are capable of describing the features for the nodes in dense networks, making RA and Jaccard achieving relatively good performance on C.elegans and USAir. The subgraph-based methods show their superior performance in link prediction. SEAL and HELP outperform other methods no matter the graph is sparse or dense. It is because that subgraph extraction selects the neighborhood that could contribute to link prediction and reduces noise in graphs. Moreover, HELP performs slightly better than SEAL with the help of HSN. On sparse networks like Router, HSN could reserve more structural information and thus makes HELP better than SEAL. It is worth noticing that PR has better link prediction ability on the five datasets than node2vec does despite that it is a similarity-based method. A possible reason is that node2vec needs a bunch of pre-defined parameters which maybe significantly influences its node representation capability. And the subsequent machine algorithms could also be another factor.

In machine learning, researchers investigate whether a method will still be useful if only given a small amount of training data. Sometimes, it is called semi-supervised learning. We would also like to compare the performances of different methods under different amount of training samples. Figure 2.6 presents the performance of the baselines as well as HELP when the ratio of test links changes from 0.2 to 0.6. In most cases, the link prediction performance drops as the decrease of training data no matter which method is applied. In dense networks like C.elegans and USAir, the performance decline of SEAL and HELP is not so significant by comparing with other methods. The smooth polylines indicate the robustness of the two subgraph-based methods. Nevertheless, AUC and AP of nearly all methods have a significant decrease on sparse networks. The drop is especially obvious when the amount of training data reduced to 60%. RA and Jaccard have poor performance on HP and Router, making the drop of AUC and AP unconspicuous. It is interesting that SEAL outperforms HELP when only 50% data are used to train the models.

2.4.6 Parameter Sensitivity

Like other deep learning methods, the performance of HELP is also affected by many factors, such as model structure and α. N_{nb}, which determines the size of subgraphs and HSNs, is the most important one among all the factors. How many neighbors does HELP need to make accurate link prediction? We investigate into the problem by varying N_{nb} from 15 to 45. Figure 2.7 shows the performance of HELP as the N_{nb} changes. In the majority cases, the performance gets improved when N_{nb} increases from 15 to 45. And when $N_{nb} = 15$, HELP could achieve comparable performance as $N_{nb} = 45$. It indicates that only part of the neighbors of the target node pair could largely contributes to link prediction. And further increasing the number of neighbors may be not so helpful. For instance, in USAir and Router, the performance of HELP becomes relatively stable when N_{nb} reaches 35. Also, the performance may become worse as N_{nb} increases, due to the introduction of useless nodes into the subgraphs in this process.

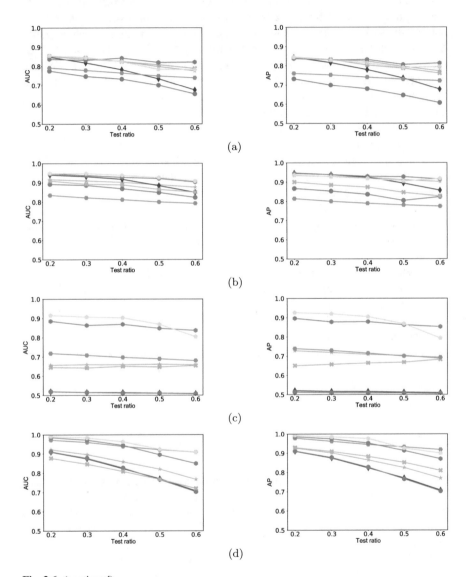

Fig. 2.6 (continued)

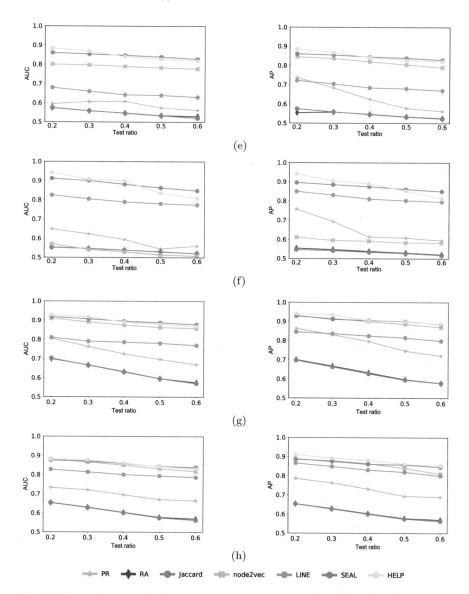

Fig. 2.6 The performance of different link prediction methods as the number of training samples changes on the five benchmark datasets. (**a**) C.elegans. (**b**) Usair. (**c**) HP. (**d**) NetScience. (**e**) Power. (**f**) Router. (**g**) Cora. (**h**) Citeseer

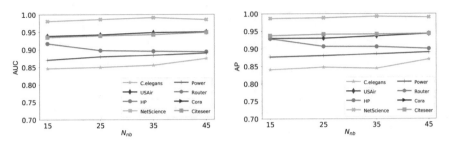

Fig. 2.7 The performance of HELP as N_{nb} changes

2.5 Conclusion

In this chapter, we present hyper-substructure enhanced link predictor (HELP) which performs link prediction over the neighborhood of given node pair. Learning from subgraphs as well as their higher-order structural information modeled by hyper-substructure network (HSN), HELP outperforms other state-of-the-art baselines, which have been proved by extensive experiments. And we also applied HELP as well as other link prediction methods on Yelp dataset, which is presented in Chapter 8. Our future research will focus on optimizing the neighborhood normalization and HSN construction process to further compress the runtime without loss of accuracy.

References

1. Atwood, J., Towsley, D.: Diffusion-convolutional neural networks. In: Advances in Neural Information Processing Systems, pp. 1993–2001 (2016)
2. Batagelj, V., Mrvar, A.: Pajek Datasets. http://vlado.fmf.uni-lj.si/pub/networks/data/mix. USAir97.net (2006)
3. Bojchevski, A., Klicpera, J., Perozzi, B., Blais, M., Kapoor, A., Lukasik, M., Günnemann, S.: Is pagerank all you need for scalable graph neural networks? In: Proceedings of the 15th MLG (2019)
4. Brin, S., Page, L.: The Anatomy of a Large-scale Hypertextual Web Search Engine (1998)
5. Cao, S., Lu, W., Xu, Q.: Deep neural networks for learning graph representations. In: AAAI, vol. 16, pp. 1145–1152 (2016)
6. Dong, Y., Chawla, N.V., Swami, A.: metapath2vec: Scalable representation learning for heterogeneous networks. In: Proceedings of the 23rd ACM SIGKDD International Conference on Knowledge Discovery and Data Mining, pp. 135–144 (2017)
7. Grover, A., Leskovec, J.: node2vec: Scalable feature learning for networks. In: Proceedings of the 22nd SIGKDD, pp. 855–864. ACM, New York (2016)
8. Hamilton, W., Ying, Z., Leskovec, J.: Inductive representation learning on large graphs. In: Proceedings of the 31st NeuralPS, pp. 1024–1034 (2017)
9. Jaccard, P.: Bulletin de la société vaudoise des sciences naturelles. Etude Comparative de la Distribution Florale dans une Portion des Alpes et des Jura **37**, 547–579 (1901)

10. Jeh, G., Widom, J.: Simrank: a measure of structural-context similarity. In: Proceedings of the Eighth ACM SIGKDD International Conference on Knowledge Discovery and Data Mining, pp. 538–543 (2002)
11. Katz, L.: A new status index derived from sociometric analysis. Psychometrika **18**(1), 39–43 (1953)
12. Kingma, D.P., Ba, J.: Adam: A method for stochastic optimization. arXiv preprint arXiv:1412.6980 (2014)
13. Kipf, T.N., Welling, M.: Variational graph auto-encoders. arXiv preprint arXiv:1611.07308 (2016)
14. Kipf, T.N., Welling, M.: Semi-supervised classification with graph convolutional networks. In: Proceedings of 5th ICLR. OpenReview.net (2017)
15. Kovács, I.A., Luck, K., Spirohn, K., Wang, Y., Pollis, C., Schlabach, S., Bian, W., Kim, D.K., Kishore, N., Hao, T., et al.: Network-based prediction of protein interactions. Nature Commun. **10**(1), 1–8 (2019)
16. Leicht, E.A., Holme, P., Newman, M.E.: Vertex similarity in networks. Phys. Rev. E **73**(2), 026120 (2006)
17. Lü, L., Zhou, T.: Link prediction in complex networks: A survey. Physica A: Stat. Mech. Appl. **390**(6), 1150–1170 (2011)
18. Mikolov, T., Chen, K., Corrado, G., Dean, J.: Efficient estimation of word representations in vector space. arXiv preprint arXiv:1301.3781 (2013)
19. Newman, M.E.: Clustering and preferential attachment in growing networks. Phys. Rev. E **64**(2), 025102 (2001)
20. Newman, M.E.: Finding community structure in networks using the eigenvectors of matrices. Phys. Rev. E **74**(3), 036104 (2006)
21. Perozzi, B., Al-Rfou, R., Skiena, S.: Deepwalk: online learning of social representations. In: Proceedings of the 20th SIGKDD, pp. 701–710 (2014)
22. Ravasz, E., Somera, A.L., Mongru, D.A., Oltvai, Z.N., Barabási, A.L.: Hierarchical organization of modularity in metabolic networks. Science **297**(5586), 1551–1555 (2002)
23. Salton, G., McGill, M.: Introduction to Modern Information Retrieval (1983)
24. Sen, P., Namata, G., Bilgic, M., Getoor, L., Galligher, B., Eliassi-Rad, T.: Collective classification in network data. AI Mag. **29**(3), 93–93 (2008)
25. Spring, N., Mahajan, R., Wetherall, D.: Measuring ISP topologies with rocketfuel. ACM SIGCOMM Comput. Commun. Rev. **32**(4), 133–145 (2002)
26. Tang, J., Qu, M., Wang, M., Zhang, M., Yan, J., Mei, Q.: Line: Large-scale information network embedding. In: Proceedings of the 24th International Conference on World Wide Web, pp. 1067–1077 (2015)
27. Veličković, P., Cucurull, G., Casanova, A., Romero, A., Liò, P., Bengio, Y.: Graph attention networks. In: Proceedings of the 6th ICLR (2018)
28. Vincent, P., Larochelle, H., Bengio, Y., Manzagol, P.A.: Extracting and composing robust features with denoising autoencoders. In: Proceedings of the 25th International Conference on Machine Learning, pp. 1096–1103 (2008)
29. Wang, D., Cui, P., Zhu, W.: Structural deep network embedding. In: Proceedings of the 22nd SIGKDD, pp. 1225–1234. ACM, New York (2016)
30. Watts, D.J., Strogatz, S.H.: Collective dynamics of small-world networks. Nature **393**(6684), 440–442 (1998)
31. Yu, W., Cheng, W., Aggarwal, C.C., Zhang, K., Chen, H., Wang, W.: Netwalk: A flexible deep embedding approach for anomaly detection in dynamic networks. In: Proceedings of the 24th ACM SIGKDD International Conference on Knowledge Discovery and Data Mining, pp. 2672–2681 (2018)
32. Zhang, M., Chen, Y.: Link prediction based on graph neural networks. In: Proceedings of 32nd NeurIPS, pp. 5171–5181 (2018)
33. Zhang, M., Cui, Z., Neumann, M., Chen, Y.: An end-to-end deep learning architecture for graph classification. In: AAAI, vol. 18, pp. 4438–4445 (2018)

34. Zhang, J., Zheng, J., Chen, J., Xuan, Q.: Hyper-substructure enhanced link predictor. In: Proceedings of the 29th ACM International Conference on Information and Knowledge Management, pp. 2305–2308 (2020)
35. Zhou, T., Lü, L., Zhang, Y.C.: Predicting missing links via local information. Eur. Phys. J. B **71**(4), 623–630 (2009)

Chapter 3
Broad Learning Based on Subgraph Networks for Graph Classification

Jinhuan Wang, Pengtao Chen, Yunyi Xie, Yalu Shan, Qi Xuan, and Guanrong Chen

Abstract Many real-world systems can be naturally represented by networks, such as biological networks, collaboration networks, software networks, social networks, etc., where subgraphs or motifs can be considered as network building blocks with particular functions to capture mesoscopic structures. Most existing studies ignored the interaction between these subgraphs, which could be of particular importance to represent the global structure at the subgraph level. In this chapter, the concept of subgraph network (SGN) is introduced and applied to network models, with algorithms designed for constructing the 1st-order and 2nd-order SGNs, which can be easily extended to build higher-order ones. Furthermore, these SGNs are used to expand the structural feature space of the underlying network, beneficial for network classification. The experiments demonstrate that the structural features of SGNs can complement that of the original network for better network classification. However, SGN model lacks diversity and is of high time-complexity, making it difficult to be widely applied in practice. Then, sampling strategies are introduced into SGNs and a novel sampling subgraph network (S^2GN) model is designed, which is scale-controllable and of higher diversity. Further, a broad learning system (BLS) is introduced into graph classification, which fully utilizes the information provided by the S^2GNs of different sampling strategies and thus can capture various aspects of the network structure more efficiently. Extensive experiments demonstrate that, by comparing with the SGN model, the S^2GN model has much lower time-complexity, which together with BLS can enhance various graph classification methods.

J. Wang · P. Chen · Y. Xie · Y. Shan · Q. Xuan (✉)
Institute of Cyberspace Security, Zhejiang University of Technology, Hangzhou, China

College of Information Engineering, Zhejiang University of Technology, Hangzhou, China
e-mail: xuanqi@zjut.edu.cn

G. Chen
Department of Electrical Engineering, City University of Hong Kong, Hong Kong SAR, China

3.1 Introduction

Studying the substructure of a large-scale network, e.g., its subgraphs, is an efficient way to understand and analyze the network. Recently, a number of studies on network subgraphs for various network applications have been reported. Ugander et al. [1] treated subgraph frequency as a local property in social networks and found that subgraph frequency can provide deep insights for identifying both social structure and graph structure in a large network. Except for subgraph frequency statistics, Benson et al. [2] developed a corresponding embedding representation through Laplacian matrix analysis method. Moreover, Wang et al. [3] designed an incremental subgraph join feature selection algorithm, which forces graph classifiers to join short-pattern subgraphs so as to generate long-pattern subgraph features. Deep learning methods for graphs achieve remarkable performance on many network analysis tasks. Yang et al. [4] proposed a NEST method, which combines the motifs and convolutional neural network. Recently, Alsentzer et al. [5] introduced a SUB-GNN to learn disentangled subgraph representations by embedding subgraphs into the GNNs, which achieves considerable performance gains on subgraph classification.

The studies mentioned above try to reveal subgraph-level patterns, which can be considered as network building blocks with particular functions, to capture mesoscopic structures. However, most of them ignored the interactions among these subgraphs, which could be of particular importance to represent the global structure at the subgraph level. In order to address this issue, Xuan et al. [6] proposed a method to establish *Subgraph Networks* (SGNs) of different orders. It is expected that such SGNs can capture the structural features in different aspects and thus may benefit the follow-up tasks, such as network classification. The details of constructing SGNs will be further discussed in Sect. 3.3. Conceptually, the SGN extracts the representative parts of the original network and then assembles them to reconstruct a new network that preserves the relationship among subgraphs. Thus, this method implicitly maintains the higher-order structures while preserving the information of local structures.

Notably, the network structure of SGN can complement the original network and the integration of SGNs' features will benefit the subsequent structure-based algorithms design and applications. However, this model can be further improved. It is observed that the rule to establish SGN is deterministic, i.e., users can generate only one SGN of each order for a network. This lack of diversity will limit the capacity of SGN to expand the latent structure space. Besides, when the number of subgraphs exceeds the number of nodes in a network, the generated SGN can be even larger than the original network, which makes it extremely time consuming to process higher-order SGNs, hindering further applications of the SGN algorithm. On the other hand, it is noted that network sampling can increase the diversity by introducing randomness, and meanwhile control the scale, providing an effective and inexpensive solution for network analysis. This feature is complementary to the SGN model. Wang et al. thus introduce *Sampling Subgraph Networks* (S^2GNs) through combining sampling strategy and SGN.

Broad Learning System (BLS) [7] is a single-layer incremental neural network, which has a good performance in training speed and classification accuracy therefore offers an alternative way of learning in deep structure. In this chapter, BLS will be adopted to fully utilize the structural information captured by S^2GN, so as to enhance the performance of graph classification. The experiments demonstrate the effectiveness of the method. The main contents of this chapter are summarized as follows:

- SGN and S^2GN are utilized to expand the structural feature space, which provides more significant and potential features for the analysis of the original network and benefits the associated algorithms.
- The broad learning system is employed for the first time to fully utilize the structural information extracted from S^2GNs generated by different sampling strategies, to enhance various graph classification algorithms based on Graph2Vec and CapsuleGNN.
- The new models are tested on three real-world network datasets, and the experimental results demonstrate that S^2GN together with BLS can indeed significantly improve the algorithm efficiency.

The rest of this chapter is organized as follows. In Sect. 3.2, some related work about subgraph networks, network representation, and the broad learning system are introduced. In Sect. 3.3, the SGN is described and the algorithms for constructing the 1st-order and 2nd-order SGNs are designed. In Sect. 3.4, three sampling strategies are developed and the construction methods of S^2GN are presented. In Sect. 3.5, BLS is used as the classifier for the classification framework. In Sect. 3.6, two feature extraction methods are introduced, and then combined with SGNs and S^2GNs, which are then applied to classify graphs in three real-world datasets. In Sect. 3.7, the computational complexities of SGNs and S^2GNs models are analyzed and compared. Finally, Sect. 3.8 concludes the chapter, with a future research outlook.

3.2 Related Work

In this section, some necessary background information is provided on subgraph networks and graph representation methods in graph mining and network science, with a brief overview of related research on the broad learning system.

3.2.1 Subgraph Networks

Subgraph is a key component in complex networks and graph mining. The structural interaction among subgraphs also plays an important role in network analysis. Subgraph network (SGN) [6] is the first model to introduce the notion of subgraph interaction, which can capture the latent high-order structural features in the original network. However, its construction is of high time-complexity. On the other hand, network sampling is an important part of network mining. Sampling methods in graph mining have two main tasks: generating node sequences for subsequent network representation [8–10] and limiting the scale of the network to simplify graphs and achieve faster graph algorithms [11, 12]. Sampling methods can simplify the network while preserving significant structural information, which is of extreme importance in graph mining. In view of this, as a variant of the SGN, sampling subgraph network (S^2GN) combined with different sampling strategies was introduced, which can enhance the performance of graph classification and reduce the time complexity of the SGN. In this chapter, SGN and S^2GN are utilized to expand the structural space for enhancing the performance of graph classification.

3.2.2 Network Representation

Network representation has received considerable attention in recent years, which allows the relational knowledge of interacting entities to be stored and accessed efficiently. The most naive network representation method is to calculate graph attributes according to certain typical topological metrics [13]. Early graph embedding methods were significantly affected by Natural Language Processing (NLP). For example, as graph-level embedding algorithms, Narayanan et al. developed Subgraph2Vec [14] and Graph2Vec [15], which achieve good performances on graph classification. Graph kernel methods [16, 17] are popular tools to capture the similarity between graphs where the kernel is equivalent to an internal product in the associated feature space. Although representing networks well, they generally have relatively high computational complexity [13], which makes it unrealizable to process large-scale networks. Graph Convolutional Networks (GCN) process the obtained information without weighting, i.e., the information of important neighbors and non-important neighbors will be put into the convolution layer in an unbiased manner. Later, Graph Attention Networks (GAT) [18] overcome this shortage by supplementing a self-attention coefficient before the convolution layer. Based on the newly proposed capsule network architecture, Zhang et al. [19] designed a CapsuleGNN to generate multiple embeddings for each graph, thereby capturing the classification-related information and the potential information with respect to the graph properties at the same time, which achieved good performance.

3.2.3 Broad Learning System

Broad Learning System (BLS) [7, 20, 21] is a single-layer incremental neural network based on the random vector function-link neural network (RVFLNN), which aims to offer an alternative way of learning in deep structure. Chen et al. [7]

showed that BLS outperforms the existing deep structure neural networks in terms of training speed. Indeed, compared with other multi-layer perceptron (MLP) training methods, BLS has a promising performance in classification accuracy and learning speed. In view of this advantages, BLS has found many applications in various fields. For example, Gao et al. [22] presented an incremental BLS for event-based object classification, which demonstrated that increasing the broad network by adding feature nodes and enhancement nodes is effective for asynchronous event-based data and provides an alternative way to deal with neuromorphic cameras. Chen et al. [23] designed a deep-broad learning system for traffic flow prediction, which increases the accuracy of traffic flow prediction, and maintains low complexity and running time. To date, BLS has been widely applied in the filed of computer vision but rarely in graph data mining. In this chapter, BLS is used for network analysis task in combination with S^2GNs. It will be shown that BLS achieves good performances in graph classification.

3.3 Subgraph Networks

To be self-contained, SGN is first reviewed, followed by the 1st-order SGN (SGN$^{(1)}$) and 2nd-order SGN (SGN$^{(2)}$) construction algorithms.

Subgraph network (SGN) maps the links in the original network into the nodes in the SGN, thereby transforming the node-level original network into a subgraph-level network.

Definition 1 (Network) Let $G(V, E)$ be an undirected network, where V and $E \subseteq (V, V)$ respectively denote the nodes and links in the network. The element $(v_i, v_j) \in E$ denotes an unordered pair of nodes v_i and v_j, i.e., $(v_i, v_j) = (v_j, v_i)$, for $i, j = 1, 2, 3, \ldots, N$, where N is the number of nodes in the network.

Definition 2 (Subgraph) For a network $G(V, E)$ and a subgraph $g_i = (V_i, E_i)$, where $g_i \subseteq G$ if and only if $V_i \subseteq V$ and $E_i \subseteq E$. Denote the sequence of subgraphs as $g = \{g_i \subseteq G | i = 1, 2, \ldots, n\}, n \leq N$.

Definition 3 (Subgraph Network) Consider an undirected network $G(V, E)$, and a SGN $G^* = f(G)$, which is a mapping from G to $G^*(V^*, E^*)$, with the nodes and links denoted by $V^* = \{g_i | j = 0, 1, \ldots, n\}$ and $E^* \subseteq (V^*, V^*)$, respectively. Two subgraphs g_i and g_j are connected if they share some common nodes or links in the original graph, i.e., $V_i \cap V_j \neq \emptyset$ or $E_i \cap E_j \neq \emptyset$. Moreover, an element $(g_i, g_j) \in E^*$ is an unordered pair of subgraphs g_i and g_j, i.e., $(g_i, g_j) = (g_j, g_i), i = 1, 2, \ldots, n$, with $n \leq N$.

According to the above definitions [6], one can actually find that SGN is derived from a higher-order mapping of the original network. Agarwal et al. [24] discussed the problem of graph representation in the domain with higher-order relations, where the node set is constructed as a p-chain, corresponding to points (0 chain), lines (1 chain), triangles (2-chain) and so on. Here, similarly, the SGN constructs the subgraphs as 1st-order, 2st-order, etc. For clarity, the following are three steps in building a SGN:

- **Extract subgraphs.** The first step is extracting subgraphs from the original network. The network has rich subgraph structures, some of which appear frequently, such as motifs [25].
- **Choose subgraph blocks.** The second step is choosing appropriate subgraph blocks. Generally, a subgraph should not be too large, otherwise the SGN may only contain a very small number of nodes, which makes subsequent analysis less significant.
- **Construct the SGN.** The final step is constructing the SGN by utilizing the subgraph blocks. After extracting enough subgraphs from the original network, the key issue is to define rules for building SGN and establish connections between these blocks. Here, for simplicity, consider two subgraphs, which are connected if they share the same node or link in the original network.

In this chapter, the most basic subgraphs (that is, lines and open triangles) are selected as subgraphs because they are simple and common in most networks.

3.3.1 First-Order SGN

From the 1st-order SGN, denoted as $SGN^{(1)}$, one can select a line namely a link as the subgraph to construct the SGN. The 1st-order SGN is also called a line graph [26].

The process of constructing $SGN^{(1)}$ from a given network is shown in Fig. 3.1. In this example, the original graph has 6 nodes connected by 6 links. First, one extracts lines as subgraphs and labels them with their corresponding terminal nodes. Then, one treats these lines as nodes in SGN, and connects them according to their labels, i.e., two lines are connected if they share the same terminal node, as shown in Fig. 3.1b. Finally, one obtains the structure of SGN with 6 nodes and 9 links as

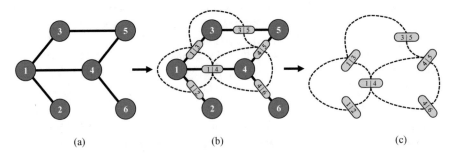

(a)   (b)   (c)

Fig. 3.1 The process to build $SGN^{(1)}$ from the original network: (**a**) the original graph, (**b**) extracting lines and establishing connection among these lines, (**c**) the structure of $SGN^{(1)}$

Algorithm 1: Constracting first-order SGN

Input: A network $G(V, E)$ with node set V and link set $E \subseteq (V \times V)$.
Output: First-order SGN, denoted by $G_1(V_1, E_1)$.
1 Initialize a node set V_1 and a link set E_1;
2 **for** *each node $u \in V$* **do**
3   Obtain the neighbors set Γ of u;
4   **for** *each node $v \in V$* **do**
5   A temporary link L = sorted pair of nodes set(u, v);
6   Regard link L as a new node in the first-order SGN;
7   Append L to node set $\widehat{V}$;
8   **end**
9   **for** *$i, j \in \widehat{V}$ and $i \neq j$* **do**
10   Append the link (i, j) to E_1;
11   **end**
12   Append $\widehat{V}$ to V_1;
13 **end**
14 **return** $G_1(V_1, E_1)$;

---

shown in Fig. 3.1c. The pseudocode for constructing SGN$^{(1)}$ is given in Algorithm 1. The input of the algorithm is the original network $G(V, E)$, and the output is the constructed SGN$^{(1)}$, denoted as $G_1(V_1, E_1)$, where V_1 and E_1 represent the nodes and links in SGN$^{(1)}$, respectively.

3.3.2 Second-Order SGN

Compared with lines, triangles can provide more insights about the local structure of the network. For example, Schiöberg et al. [27] studied the evolution of triangles in the Google+ online social network, where some valuable information was obtained during the appearance and pruning of various triangles.

Now, construct a higher-order subgraph by considering the connection pattern between three nodes. Compared with two nodes, the connection mode between three nodes is more diverse. Here, only the connected subgraphs are considered, and the subgraphs with less than two links are ignored. Here, the open triangle is defined as a subgraph to establish a 2nd-order SGN, denoted by SGN$^{(2)}$. Second order means that there are two links in each open triangle, and if two open triangles share the same link, the two open triangles are connected in SGN$^{(2)}$. Note that the same link instead of the same node is used here to avoid getting a very dense SGN$^{(2)}$. This is because, usually, dense networks with higher connection probability of each node pair tend to provide less discriminative information for local structures.

The construction process from SGN$^{(1)}$ to SGN$^{(2)}$ is illustrated by Fig. 3.2. In the line graph SGN$^{(1)}$, further extract lines to obtain hollow triangles as subgraphs, and mark them with the corresponding three nodes, as shown in Fig. 3.2b. Finally, an SGN$^{(2)}$ with 8 nodes and 15 links is obtained, as shown in Fig. 3.2c. The pseudocode

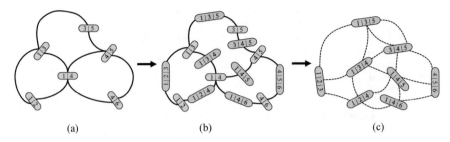

Fig. 3.2 The process to build SGN$^{(2)}$ from the first-order subgraph network: (**a**) SGN$^{(1)}$ in Fig. 3.1, (**b**) extracting lines and establishing connections among these lines, (**c**) the structure of SGN$^{(2)}$

Algorithm 2: Constracting second-order SGN

Input: A network $G(V, E)$ with node set V and link set $E \subseteq (V \times V)$.
Output: Second-order SGN, denoted by $G_2(V_2, E_2)$.
1 Initialize a node set V_2 and a link set E_2;
2 for *each node $u \in V$* **do**
3   Obtain the neighbors set Γ of u;
4   The set of all node pairs in the neighbor collection $\widehat{\Gamma}$;
5   **for** *each node pair $(v_1, v_2) \in \widehat{\Gamma}$* **do**
6   A temporary link L = sorted pair of nodes set(u, v_1, v_2);
7   Regard link L as a new node in the first-order SGN;
8   Append L to node set $\widehat{V}$;
9   **end**
10   **for** *$i, j \in \widehat{V}$ and $i \neq j$* **do**
11   Append the link (i, j) to E_2;
12   **end**
13   Append $\widehat{V}$ to V_2;
14 end
15 return $G_2(V_2, E_2)$;

---

for constructing SGN$^{(2)}$ is given in Algorithm 2. The input of the algorithm is the original network $G(V, E)$, and the output is the constructed SGN$^{(2)}$, represented by $G_2(V_2, E_2)$, where V_2 and E_2 represent the node and link sets in SGN$^{(2)}$, respectively.

As SGN gradually maps to the higher-order network, one can obtain more and richer feature information. SGN$^{(1)}$ can reveal the topological interaction between the links of the original network. Fu et al. [28] adopted SGN to predict the link weights of given networks. SGN$^{(2)}$ is obtained by further iterative mapping based on SGN$^{(1)}$, so the second-order information of the nodes can be captured. Higher-order SGNs will contain more hidden information, but these hidden information may play a smaller role in subsequent applications. Therefore, here the focus is on the SGNs of the first two orders.

3.4 Sampling Subgraph Networks

Next, the S²GN is reviewed. S²GN is proposed as a variant of SGN through introducing sampling strategies into the SGN algorithm. In this section, several sampling strategies are described, and the construction of S²GN is discussed.

3.4.1 Sampling Strategies

Network sampling can simplify a graph while preserving its significant structural information, which is of extreme importance in graph data mining. Here, three sampling algorithms are reviewed: biased walk, spanning tree and forest fire.

3.4.1.1 Biased Walk (BW)

Biased walk sampling is a very common sampling strategy. Here, the walking mechanism of Node2Vec [29] is adopted to preserve the homogeneity and structure of nodes by integrating depth-first search (DFS) and breadth-first search (BFS) (as shown in Fig. 3.3a). In the mechanism of Node2Vec, it defines a 2nd-order random walk, which is guided by the two parameters p and q (as shown in Fig. 3.3b). To start, assume that one walks from node A to node B and then needs to determine the next step. The transition probability P from node B to node C is then defined as

$$P_{(B,C)} = f_{pq}(A, C) = \begin{cases} \frac{1}{p}, d_{(A,C)} = 0 \\ 1, d_{(A,C)} = 1 \\ \frac{1}{q}, d_{(A,C)} = 2 \end{cases}$$

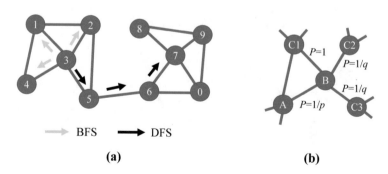

(a)   (b)

Fig. 3.3 (a) BFS and DFS walk strategies from node 3. (b) Illustration of evaluating the next step out of node B. Edge labels indicate search biases P

where $d_{A,C}$ represents the shortest distance between nodes A and C, and its value must be in $\{0, 1, 2\}$. The pseudocode for biased walk is given in Algorithm 3.

Algorithm 3: Biased walk sampling

Input: A network $G(V, E)$ with node set V and link set $E \subseteq (V \times V)$.
Output: The substructure $G_b(V_b, E_b)$.
1 Initialize a source node $v_0 \in V$, added into V_b;
2 Append v_0 to V_b;
3 **while** *walk length L* **do**
4 $\quad$ $v = V_b[-1]$;
5 $\quad$ $V_v = \text{GetNeighbors}(v, G)$;
6 $\quad$ $v_{next} = \text{AliasSample}(V_v, \pi)$;
7 $\quad$ Append v_{next} to V_b;
8 $\quad$ Append (v, v_{next}) to E_b;
9 **end**
10 **return** $G_b(V_b, E_b)$;

3.4.1.2 Spanning Tree (ST)

A spanning tree [30] is defined as a subgraph of a connected tree graph G, which connects all the nodes together with minimum possible number of edges. Here, since the datasets used are all unweighted networks, the largest spanning tree is equivalent to the smallest spanning tree. In this chapter, the classical Kruskal algorithm [31] is used to generate spanning trees (as shown in Fig. 3.4) and the weight values of the links are all set to one.

The Kruskal algorithm is a greedy algorithm for generating spanning tree as follow:

- Step 1. Create a set of trees $\mathscr{F}$ where each node in the graph belongs to a tree;
- Step 2. Create a set of edges $\mathscr{E}$ including all edges of the graph;

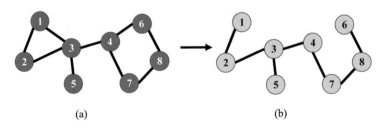

(a)   (b)

Fig. 3.4 Obtaining a spanning tree. (**a**) Original network. (**b**) Spanning tree structure

- Step 3. While $\mathscr{E} \neq \emptyset$ and $|\mathscr{F}| \neq 1$

  - select any edge from $\mathscr{E}$;
  - if this edge connects two different trees, then combine it with the two trees to generate a new tree and add it into $\mathscr{F}$;
  - otherwise discard this edge;

- Step 4. The $\mathscr{F}$ has a minimum spanning tree at the termination of the iteration.

3.4.1.3 Forest Fire (FF)

Forest fire sampling was first proposed by Leskovec and Faloutsos in 2006 [11]. Here, a specific algorithm is introduced for forest fire sampling. Given a network, first randomly select a node v_0, and then generate a random number X, following the geometric distribution with mean $p_f/(1 - p_f)$. Here, the parameter p_f is called the forward burning probability (set to 0.2 [11]). Then, X edges with node v_0 as terminal node will be selected, where the another terminal node of each edge has not been visited. One can use any method of generating random numbers to find the unvisited nodes with the current burning nodes as source nodes, one by one, until enough nodes are burned. To avoid duplication, nodes cannot be visited twice during the forest fire sampling method. If the fire dies, select a node randomly to restart again. The pseudocode for forest fire sampling is given in Algorithm 4.

Algorithm 4: Forest fire sampling

Input: A network $G(V, E)$ with node set V and link set $E \subseteq (V \times V)$, the forward burning probability p_f.
Output: The substructure $G_s(V_s, E_s)$.
1 Initialize a neighbor list N and a temporary variable $G_s(V_s, E_s)$;
2 Randomly choose the first node v_0;
3 Generate X following the geometric distribution with mean $p_f/(1 - p_f)$;
4 $n =$ The number of v_0' neighbor;
5 **if** $X \leq n$ **then**
6 $\quad$ N$\leftarrow$ Sort the neighbors of v_0 according to the degree and choose the top X neighbors;
7 **end**
8 **for** *node T in N* **do**
9 $\quad$ **if** T *in* V_s **then**
10 $\quad\quad$ continue;
11 $\quad$ **else**
12 $\quad\quad$ Append the node T to V_s;
13 $\quad\quad$ Append the link (T, v_0) to E_s;
14 $\quad\quad$ Forest fire recursive function(G, p_f, T);
15 $\quad$ **end**
16 **end**
17 **return** $G_s(V_s, E_s)$;

Using any of the above three sampling strategies, one can map the original network into many substructures. As a result, more characteristic information in the network can be abstracted, which also provides favorable preconditions for downstream algorithms.

3.4.2   Construction of S^2GN

Most networks in the real world have complex structures. Therefore, the generated SGNs are typically of large-scale and even denser than the original networks. This will not only reduce the efficiency of the algorithms, but also introduce some "noise" into structures, which will reduce the accuracy of the algorithms. In view of this, the original SGN model is optimised to construct S^2GN. In particular, multiple sampling strategies are introduced, filtering the original complex network as substructures, thereby establishing a new SGN. The pseudocode for constructing S^2GN is given in Algorithm 5. Generally, the S^2GN algorithm is divided into three parts: selecting source node, sampling substructure and constructing subgraph network. The algorithm steps are described as follows:

- **Selecting source node.** There are many ways to select the source node: (i) randomly selecting a node as the source node; (ii) selecting an initial node according to the importance of the node. In this chapter, the second method is adopted to better capture the key structure of a network.
- **Sampling substructures.** After determining the initial source node, a substructure can be obtained by performing a certain sampling strategy to extract the main context of the current network. According to different sampling strategies, various sampling substructures can be generated, and the rich structural information in the current network can be obtained.
- **Constructing S^2GN.** The sampling substructure is used as input to construct a subgraph network. Note that sampling and SGN construction are repeated interactively, so as to obtain higher-order S^2GNs. That is, each time the current network is mapped to a subgraph network, and then a sampling operation is performed to obtain various relatively simple sampling substructures.

3.5   BLS Classifier Based on S^2GN

3.5.1   BLS Classifier

BLS [7, 20] is proposed as an alternative method of deep learning network. It is designed such that mapping features are input into RVFLNN. As shown in Fig. 3.5, a specific illustration of BLS is given, which will be used as the classifier of the network.

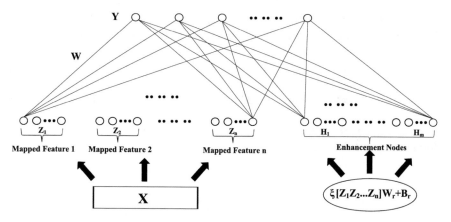

Fig. 3.5 The framework of BLS classifier

Algorithm 5: Constracting sampling subgraph network

Input: A network $G(V, E)$ with node set V and link set $E \subseteq (V \times V)$, the order of SGN T, sampling strategy $f_s(\cdot)$, sampling walks L.
Output: Sampling subgraph network, denoted by $G'(V', E')$.
1 Initialize a temporary variable $G' = G$;
2 **while** T **do**
3       $G' = \text{GetMaxSubstracture}(G')$;
4   Source node $o = \text{NodeRanking}(V')$;
5   Initial sampling $v_i = o, W_v \leftarrow [o], W_e \leftarrow \emptyset$;
6   **for** $i = 1$ *to* $L - 1$ **do**
7   Sampling link $e_i = f_s(v_i)$;
8   Update current node $v_i = dst(e_i)$;
9   Append v_i to W_v, e_i to W_e;
10   **end**
11      $V_s \leftarrow W_v$, $E_s \leftarrow W_e$;
12      $G_{sgn} = \text{SGN Algorithms}(G_s)$;
13      $G' \leftarrow \text{Relabel}(G_{sgn})$;
14      $T = T - 1$;
15 **end**
16 **return** $G'(V', E')$;

In this process, we treat the training features as the graph input X, and then transform X into n random feature spaces by feature mapping:

$$Z_i = \phi(X W_{z_i} + \beta_{z_i}), i = 1 \ldots n \tag{3.1}$$

where the weights W_{z_i} and the bias term β_{z_i} are generated randomly with appropriate dimensions, n is the number of groups of mapped features, and $\phi(\cdot)$ indicates the linear mapping.

Here, denote the feature space of training samples by $Z^n = [Z_1, Z_2, \ldots, Z_n]$. Then, the jth group of enhancement nodes is defined by

$$H_j = \xi(Z^n W_{r_j} + B_{r_j}), j = 1, 2, \ldots, m \tag{3.2}$$

where W_{r_j} is the enhancement weight, B_{r_j} is the bias term, and $\xi(\cdot)$ is a nonlinear activation function.

Similarly, denote the enhancement layer by $H^m = [H_1, H_2, \ldots, H_m]$. Thus, the form of output $\hat{Y}$ is as follows:

$$\hat{Y} = [Z^n, H^m]W = AW \tag{3.3}$$

where $A = [Z^n, H^m]$ are the features, combining the enhancement nodes and feature nodes, and W is the weight matrix that connects the feature nodes and enhancement nodes to the output layer. The W should be optimized by

$$min_W ||Y - AW||_2^2 + \lambda ||W||_2^2 \tag{3.4}$$

where λ is a regularization coefficient. Then, through a simple equivalent transformation [7], one finally gets the following formula:

$$W = (A^T A + \lambda I)^{-1} A^T Y \tag{3.5}$$

Now, one has the trained model with the weight matrix W, therefore can test its performance using the rest of the dataset. In fact, there are several hyperparameters in this model, such as the number of feature nodes k, the number of enhancement nodes m, the regularization coefficient λ and the shrink coefficient s of $\xi(\cdot)$. In the experiment, the regularization coefficient λ is set to 2^{-10} and shrink coefficient s to 0.8. For MUTAG, PTC, and PROTEINS, set the number of feature nodes k as 50, 80, and 100, and the number of enhancement nodes m as 40, 60, and 80, respectively. For different samples, the optimal parameters are set near the above parameters.

3.5.2 Classification Framework

According to the above setting, one can design a framework for graph classification by combining S^2GN with the BLS classifier, as shown in Fig. 3.6.

To begin with, construct three S^2GNs according to the method described in Sect. 3.4.2, i.e., $S^2GN^{(0)}$, $S^2GN^{(1)}$ and $S^2GN^{(2)}$, and then map them into different feature spaces. By graph feature extraction methods, one can get the feature representations of S^2GNs and then fusion them into $X = [X_0||X_1||X_2]$, where $[a||b]$ means to merge vector a and vector b. The fused vector X can be regarded as the input of the BLS classifier as shown in Fig. 3.5. Here, the same sampling strategy and feature extraction method are adopted for conducting classification. Since the

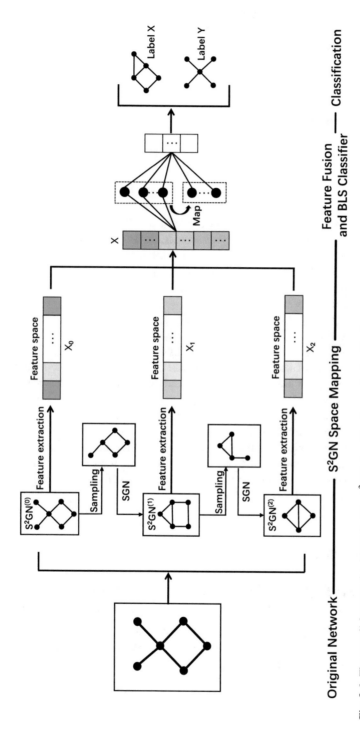

Fig. 3.6 The overall framework with BLS of the S^2GN algorithm for graph classification

same original network can generate various S^2GNs by different sampling strategies, X contains abundant structural information of different aspects. Therefore, this framework combining S^2GNs with BLS can enhance the performance of the original network classification.

3.6   Experiment

3.6.1   Graph Classification

The performance of the above framework on graph classification is evaluated by simulation. Graph classification is one of the most important data mining tasks, which has been widely used in the field of biochemistry, such as protein classification and molecular toxicity classification. Typically, graph classification focuses on transforming discrete graphs into numerical features. One can use some machine learning algorithms to effectively classify various graphs.

Consider graph $G = (V, E)$ from the set $\mathscr{G} = \{G_i\}$, where $i = 1, 2, 3 \ldots, N$. The node and edge collections are $V = \{v_1, v_2, v_3, \ldots, v_n\}$ and $E = \{e_1, e_2, e_3, \ldots, e_n\} \subseteq (V \times V)$, respectively. Each graph G has a corresponding category label $y \in \mathscr{C}$, where $\mathscr{C} = \{1, 2, 3, \ldots, k\}$ is the set containing k different labels. The aim of graph classification is to find a mapping function $f : \mathscr{G} \to \mathscr{C}$ to predict the label of each graph in $\mathscr{G}$. Generally, one can train the model using the training set with known category labels and evaluate its performance using the test set with unknown labels. By comparing the output $\widehat{y} = f(G)$ with the true test label y, one can evaluate the classification algorithm by accuracy.

To date, a large number of graph classification methods have been proposed, like graph embedding method Graph2Vec and deep learning method CapsuleGNN. In this section, a brief introduction of these two network representation methods is given, and then some experiments on three datasets are performed for evaluating the performances of both SGN and S^2GN models.

3.6.2   Datasets

Three datasets, MUTAG, PTC and PROTEINS, will be used for graph classification experiments. The basic statistics of these datasets are presented in Table 3.1.

- MUTAG [32] dataset is about heteroaromatic nitro and mutagenic aromatic compounds, with nodes and links representing atoms and the chemical bonds between them, respectively. They are labeled according to whether there is a mutagenic effect on a special bacteria.
- PTC [33] dataset includes 344 chemical compound graphs whose labels are determined by their carcinogenicity for rats.

Table 3.1 Basic statistics of the 3 datasets. #Graphs is the number of graphs. #Classes is the number of classes. #Positive and #Negative are the numbers of graphs in the two different classes

Dataset	#Graphs	#Classses	#Positive	#Negative
MUTAG	188	2	125	63
PTC	344	2	152	192
PROTEINS	1113	2	663	450

- PROTEINS [34] dataset comprises of 1113 graphs. The nodes are Secondary Structure Elements (SSEs) and the links are neighbors in the amino-acid sequence or in the 3D space. These graphs represent either enzyme or non-enzyme proteins.

3.6.3 Network Representation

Network representation is a method of mapping graphs into vectors while retaining as many topological features as possible. Here, two network representation methods are used to extract graph features, including graph embedding method Graph2Vec and deep learning method CapsuleGNN.

- **Graph2Vec**: This is the first unsupervised embedding approach for an entire network, which is based on the extending word-and-document embedding technique that has shown great advantages in NLP. Graph2Vec establishes the relationship between a network and the rooted subgraphs using a similar model to Doc2Vec. Graph2Vec first extracts rooted subgraphs and provides corresponding labels into the vocabulary, and then trains a skipgram model to obtain a representation of the entire network.
- **CapsuleGNN**: This method is inspired by CapsNet [35], which utilizes the concept of capsules to overcome the shortcoming of existing GNN-based graph representation algorithms. CapsuleGNN extracts node features in the form of capsules and uses routing mechanisms to capture important information at the graph level. The model generates multiple embeddings for each graph in order to capture graph properties from different aspects.

For *Graph2Vec*, the embedding dimension is adopted following [15]. Graph2vec is based on the rooted subgraphs adopted in the WL kernel. The parameter height of the WL kernel is set to 3. Since the embedding dimension is predominant for learning performances, a commonly-used value of 1024 is adopted. The other parameters are set to defaults: the learning rate is set to 0.5, the batch size is set to 512 and the epochs is set to 1000. The default parameters are used for *CapsuleGNN* and the multiple embeddings of each graph are flattened as the input.

3.6.4 SGN for Graph Classification

As described in Sect. 3.3, the proposed SGNs can be used to expand structural feature spaces. In order to study the effectiveness of the 1st-order and 2nd-order SGNs, i.e., $SGN^{(1)}$ and $SGN^{(2)}$, the classification results were compared based on different numbers of networks, i.e., $SGN^{(0)}$, $SGN^{(1)}$, $SGN^{(2)}$ $SGN^{(0,1)}$, $SGN^{(0,2)}$ and $SGN^{(0,1,2)}$. Without loss of generality, the well-known logistic regression is chosen as the classification model. Meanwhile, for each feature extraction method, the feature space is first expanded by using SGNs, and then the dimension of the feature vectors is reduced to the same value as that of the feature vector obtained from the original network using PCA in the experiments, for a fair comparison. Each dataset is randomly split into ninefolds for training and onefold for testing. Here, the F_1-*Score* is adopted as the metric to evaluate the classification performance:

$$\mathscr{F} = \frac{2\mathscr{P}\mathscr{R}}{\mathscr{P} + \mathscr{R}}, \tag{3.6}$$

and the Gain can be calculated by

$$Gain = \frac{\mathscr{F}^{(0,1,2)} - \mathscr{F}^{(0)}}{\mathscr{F}^{(0)}} \times 100\% \tag{3.7}$$

where $\mathscr{P}$ and $\mathscr{R}$ are the precision and recall, respectively. To exclude the random effect of the fold assignment, experiment is repeated for 500 times and then the average F_1-*Score* and its standard deviation are recorded. The results for Graph2Vec and CapsuleGNN methods are shown in Tables 3.2 and 3.3, respectively.

From Tables 3.2 and 3.3, one can see that the original network appears to provide more structural information. The classification model based on $SGN^{(0)}$ performs better, with a higher F_1-*Score*, than those based on $SGN^{(1)}$ or $SGN^{(2)}$. This is reasonable because there is information loss in the processes of constructing SGNs. More interestingly, the performance of the classification model based on two networks i.e., $SGN^{(0,1)}$ and $SGN^{(0,2)}$, is better than the classification model based on

Table 3.2 Classification results on MUTAG, PTC and PROTEINS, in terms of F_1-*Score*, based on Graph2Vec method with the combinations of SGNs of different orders, the bold values are the best results

Dataset	MUTAG	PTC	PROTEINS
Original	83.15 ± 9.25	60.17 ± 6.86	73.30 ± 2.05
$SGN^{(1)}$	63.16 ± 4.68	56.80 ± 5.39	60.27 ± 2.05
$SGN^{(2)}$	68.95 ± 8.47	57.35 ± 3.83	59.82 ± 4.11
$SGN^{(0,1)}$	83.42 ± 5.40	59.03 ± 3.36	74.12 ± 1.57
$SGN^{(0,2)}$	81.32 ± 3.80	61.76 ± 3.73	73.09 ± 1.28
$SGN^{(0,1,2)}$	$\mathbf{86.84 \pm 5.70}$	$\mathbf{63.24 \pm 6.70}$	$\mathbf{74.44 \pm 3.09}$
Gain	4.44%	5.10%	1.56%

Table 3.3 Classification results on MUTAG, PTC and PROTEINS, in terms of F_1-$Score$, based on CapsuleGNN method with the combinations of SGNs of different orders, the bold values are the best results

Dataset	MUTAG	PTC	PROTEINS
Original	86.32 ± 7.52	62.06 ± 4.25	75.89 ± 3.51
SGN$^{(1)}$	83.68 ± 8.95	61.76 ± 5.00	74.64 ± 3.55
SGN$^{(2)}$	82.63 ± 7.08	58.82 ± 3.95	73.39 ± 6.03
SGN$^{(0,1)}$	87.37 ± 8.55	63.53 ± 4.40	76.25 ± 3.53
SGN$^{(0,2)}$	87.89 ± 5.29	62.20 ± 6.14	73.00 ± 3.17
SGN$^{(0,1,2)}$	$\mathbf{89.47 \pm 7.44}$	$\mathbf{64.12 \pm 3.67}$	$\mathbf{76.34 \pm 4.13}$
Gain	3.65%	2.19%	0.59%

a single network in most cases, which prove that SGNs can indeed provide the latent and significant structural information. Furthermore, when the three single networks are considered together, i.e., SGN$^{(0,1,2)}$, they can achieve the best performance in the classification.

3.6.5 S²GN for Graph Classification

In this experiment, the network in the dataset is divided to ten equal parts, two of which are selected as the test set, and the remaining eight are used as the training set. The above algorithms are used to generate S²GNs of different orders, and then Graph2Vec and CapsuleGNN methods are adopted to learn the feature representations of S²GNs. Finally, the BLS method is applied for classification and to calculate the F1-Score. In order to avoid accidental sampling, each sampling strategy was carried out 10 sampling averaging. Based on Graph2Vec and CapsuleGNN methods, the obtained experimental results are shown in Tables 3.4 and 3.5, respectively.

The influence of each sampling method on the model is compared between the two feature extraction methods. It is found that, under the same feature extraction method, each sampling strategy has different advantages and disadvantages, which may be related to the specific network structure in the dataset. With the two feature extraction algorithms, the dataset MUTAG has the best classification effect under biased walk. Here, the results based on three sampling methods are compared with those based on the original method. It is found that the improved sampling subgraph network algorithm can maintain the original accuracy and even outperform

Table 3.4 Classification results of three sampling strategies in MUTAG, PTC and PROTEINS, in terms of F_1-$Score$, based on Graph2Vec method, the bold values are the best results

Dataset	MUTAG	PTC	PROTEINS
Original	83.15 ± 9.25	60.17 ± 6.86	73.30 ± 2.05
S²GN-BW	$\mathbf{86.80 \pm 5.02}$	62.90 ± 2.19	75.44 ± 3.85
S²GN-ST	82.03 ± 3.76	62.39 ± 6.36	74.33 ± 2.86
S²GN-FF	83.33 ± 6.13	62.46 ± 5.17	73.77 ± 2.15
BLS-S²GN	83.63 ± 6.84	$\mathbf{63.28 \pm 6.06}$	$\mathbf{74.92 \pm 2.58}$
Gain	0.58%	5.17%	2.21%

Table 3.5 Classification results of three sampling strategies in MUTAG, PTC and PROTEINS, in terms of F_1-$Score$, based on CapsuleGNN method, the bold values are the best results

Dataset	MUTAG	PTC	PROTEINS
Original	86.32 ± 7.52	62.06 ± 4.25	75.89 ± 3.51
S^{2}GN-BW	**91.84 ± 5.00**	66.06 ± 3.34	77.47 ± 2.35
S^{2}GN-ST	89.21 ± 5.05	63.50 ± 3.71	76.85 ± 2.54
S^{2}GN-FF	86.10 ± 5.32	65.77 ± 4.42	76.37 ± 1.90
BLS-S^{2}GN	91.63 ± 2.89	**66.10 ± 6.37**	**78.32 ± 2.94**
Gain	6.39%	6.50%	3.20%

the optimal result of the original algorithm. Experiments show that the variance of all accuracy on the dataset is reduced. For example, under the condition of CapsuleGNN feature extraction for the MUTAG dataset, the highest average accuracy is around 86.32%, and the variance is 0.0752. While using BLS-S^{2}GN, the highest average accuracy can reach 91.84%, and the variance is reduced to 0.0289. All cases show that the newly proposed classification model can be combined with different feature extraction methods, which can further improve the accuracy and stability of classification.

3.7 Computational Complexity

Now, it is to analyze the computational complexity in building SGNs. Denote by $|V|$ and $|E|$ the numbers of nodes and links, respectively, in the original network. The average degree of the network is calculated by

$$K = \frac{1}{|V|} \sum_{i=1}^{|V|} k_i = \frac{2|E|}{|V|}, \tag{3.8}$$

where k_i is the degree of node v_i. Based on Algorithm 1, the time complexity in transforming the original network to SGN$^{(1)}$ is

$$\mathscr{T}_1 = \mathscr{O}(K|V| + |E|^2) = \mathscr{O}(|E|^2 + |E|) = \mathscr{O}(|E|^2). \tag{3.9}$$

Then, the number of nodes in SGN$^{(1)}$ is equal to $|E|$ and the number of links is $\sum_{i=1}^{|V|} k_i^2 - |E| \leq |E|^2 - |E|$ [26]. Similarly, one can get the time complexity in transforming SGN$^{(1)}$ to SGN$^{(2)}$, as

$$\mathscr{T}_2 \leq \mathscr{O}((|E|^2 - |E|)^2) = \mathscr{O}(|E|^4). \tag{3.10}$$

Meanwhile, the computational complexity of these methods is evaluated in terms of the average computational time of SGN and S^{2}GN generated by the three sampling strategies on the three datasets. The results are presented in Table 3.6, where one can see that, overall, the computational time of S^{2}GN is much less than

Table 3.6 Average computational time to establish SGN and S^{2}GNs by the three sampling strategies on the three datasets

Time (Seconds)	SGN	S^{2}GN		
		BW	ST	FF
MUTAG	1.58×10^2	0.252	0.090	0.382
PTC	1.93×10^3	0.804	0.607	0.985
PROTEINS	3.20×10^3	1.161	1.625	3.697

that of SGN for each sampling strategy on each dataset, decreasing from hundreds of seconds to less than 4 s. These results suggest that, by comparing with SGN, the S^{2}GN model can indeed largely increase the efficiency of the network algorithms.

In fact, it is possible to estimate the time complexity of S^{2}GN model in theory. For biased walk, consider the 2nd random walk mechanism of Node2Vec, where each step of a random walk is based on the transition probability α, which can be precomputed, so the time consumption of each step using alias sampling is $\mathcal{O}(1)$. The Kruskal algorithm used to generate spanning trees is a greedy algorithm, which has $\mathcal{O}(|E|log(|E|))$ time complexity. Fire forest is an exploration-based method. The difference between this method and the random walk method is that, when a node is visited, it will no longer be visited again in the fire forest. It is known that the computational complexity of SGN$^{(1)}$ is $\mathcal{O}(|E|^2)$ and that of constructing SGN$^{(2)}$ is $\mathcal{O}(|E|^4)$. The S^{2}GN model constrains the expansion of the network scale and reduces the cost of constructing SGNs to the fixed $\mathcal{O}(|E|^2)$. Thus, the time computational complexity $\mathcal{T}$ of the S^{2}GN model can be calculated as

$$\mathcal{T} \leq \mathcal{O}(|E|log|E| + |E|^2|) \tag{3.11}$$

Combining with the different sampling strategies, one can see that the time complexity of S^{2}GN is much lower than that of SGN.

3.8 Conclusion

In this chapter, after reviewing the notions of SGN and S^{2}GN, the BLS is introduced to graph data mining. Moreover, a classification framework is introduced that combines S^{2}GN, feature representation, and BLS to enhance the performance of graph classification task. SGN and S^{2}GN can generate different higher-order graphs to capture latent structural information of the original network from various aspects and expand the feature space. Experiments on three datasets demonstrate that BLS can fully utilize these latent features to achieve significant improvement in graph classification. In addition, it is found that, compared with SGN, S^{2}GN has much lower time complexity, which was reduced by almost two orders of magnitude. More significantly, combined with BLS, it has competitive performances on graph classification.

References

1. Ugander, J., Backstrom, L., Kleinberg, J.: Subgraph frequencies: mapping the empirical and extremal geography of large graph collections. In: Proceedings of the 22nd International Conference on World Wide Web, pp. 1307–1318. ACM, New York (2013)
2. Benson, A.R., Gleich, D.F., Leskovec, J.: Higher-order organization of complex networks. Science **353**(6295), 163–166 (2016)
3. Wang, H., Zhang, P., Zhu, X., Tsang, I.W.-H., Chen, L., Zhang, C., Wu, X.: Incremental subgraph feature selection for graph classification. IEEE Trans. Know. Data Eng. **29**(1), 128–142 (2017)
4. Yang, C., Liu, M., Zheng, V.W., Han, J.: Node, motif and subgraph: leveraging network functional blocks through structural convolution. In: 2018 IEEE/ACM International Conference on Advances in Social Networks Analysis and Mining (ASONAM), pp. 47–52. IEEE, Piscataway (2018)
5. Alsentzer, E., Finlayson, S.G., Li, M.M., Zitnik, M.: Subgraph neural networks. Preprint. arXiv:2006.10538 (2020)
6. Xuan, Q., Wang, J., Zhao, M., Yuan, J., Fu, C., Ruan, Z., Chen, G.: Subgraph networks with application to structural feature space expansion. IEEE Trans. Knowl. Data Eng. **33**(6), 2776–2789 (2021)
7. Philip Chen, C.L., Liu, Z.: Broad learning system: an effective and efficient incremental learning system without the need for deep architecture. IEEE Trans. Neural Netw. Learn. Syst. **29**(1):10–24 (2017)
8. Kurant, M., Markopoulou, A., Thiran, P.: On the bias of BFS (breadth first search). In: 2010 22nd International Teletraffic Congress (ITC 22), pp. 1–8. IEEE, Piscataway (2010)
9. Perozzi, B., Al-Rfou, R., Skiena, S.: Deepwalk: online learning of social representations. In: Proceedings of the 20th ACM SIGKDD International Conference on Knowledge Discovery and Data Mining, pp. 701–710 (2014)
10. Tang, J., Qu, M., Wang, M., Zhang, M., Yan, J., Mei, Q.: Line: large-scale information network embedding. In: Proceedings of the 24th International Conference on World Wide Web, pp. 1067–1077 (2015)
11. Leskovec, J., Faloutsos, C.: Sampling from large graphs. In: Proceedings of the 12th ACM SIGKDD International Conference on Knowledge Discovery and Data Mining, pp. 631–636 (2006)
12. Satuluri, V., Parthasarathy, S., Ruan, Y.: Local graph sparsification for scalable clustering. In: Proceedings of the 2011 ACM SIGMOD International Conference on Management of Data, SIGMOD '11, New York, pp. 721732 (2011)
13. Li, G., Semerci, M., Yener, B., Zaki, M.J.: Graph classification via topological and label attributes. In: Proceedings of the 9th International Workshop on Mining and Learning with Graphs (MLG), San Diego, USA, vol. 2 (2011)
14. Narayanan, A., Chandramohan, M., Chen, L., Liu, Y., Saminathan, S.: subgraph2vec: learning distributed representations of rooted sub-graphs from large graphs. In: International Workshop on Mining and Learning with Graphs (2016)
15. Narayanan, A., Chandramohan, M., Venkatesan, R., Chen, L., Liu, Y., Jaiswal, S.: graph2vec: learning distributed representations of graphs. In: International Workshop on Mining and Learning with Graphs (2017)
16. Shervashidze, N., Schweitzer, P., van Leeuwen, E.J., Mehlhorn, K., Borgwardt, K.M.: Weisfeiler-Lehman graph kernels. J. Mach. Learn. Res. **12**, 25392561 (2011)
17. Yanardag, P., Vishwanathan, S.V.N.: Deep graph kernels. In: Proceedings of the 21th ACM SIGKDD International Conference on Knowledge Discovery and Data Mining, pp. 1365–1374. ACM, New York (2015)
18. Velikovi, P., Cucurull, G., Casanova, A., Romero, A., Li, P., Bengio, Y.: Graph attention networks. In: International Conference on Learning Representations (2018)

19. Zhang, X., Chen, L.: Capsule graph neural network. In: International Conference on Learning Representations (2019)
20. Philip Chen, C.L., Liu, Z.: Broad learning system: a new learning paradigm and system without going deep. In: 2017 32nd Youth Academic Annual Conference of Chinese Association of Automation (YAC), pp. 1271–1276. IEEE, Piscataway (2017)
21. Jin, J.-W., Philip Chen, C.L.: Regularized robust broad learning system for uncertain data modeling. Neurocomputing **322**, 58–69 (2018)
22. Gao, S., Guo, G., Philip Chen, C.L.: Event-based incremental broad learning system for object classification. In: Proceedings of the IEEE International Conference on Computer Vision Workshops, pp. 2989–2998 (2019)
23. Chen, M., Wei, X., Gao, Y., Huang, L., Chen, M., Kang, B.: Deep-broad learning system for traffic flow prediction toward 5g cellular wireless network. In: 2020 International Wireless Communications and Mobile Computing (IWCMC), pp. 940–945. IEEE, Piscataway (2020)
24. Agarwal, S., Branson, K., Belongie, S.: Higher order learning with graphs. In: Proceedings of the 23rd International Conference on Machine Learning, pp. 17–24. ACM, New York (2006)
25. Wernicke, S.: Efficient detection of network motifs. IEEE/ACM Trans. Comput. Biol. Bioinform. (TCBB) **3**(4), 347–359 (2006)
26. Harary, F., Norman, R.Z.: Some properties of line digraphs. Rendiconti del Circolo Matematico di Palermo **9**(2), 161–168 (1960)
27. Schiöberg, D., Schneider, F., Schmid, S., Uhlig, S., Feldmann, A.: Evolution of directed triangle motifs in the google+ OSN. Preprint. arXiv:1502.04321 (2015)
28. Fu, C., Zhao, M., Fan, L., Chen, X., Chen, J., Wu, Z., Xia, Y., Xuan, Q.: Link weight prediction using supervised learning methods and its application to Yelp layered network. IEEE Trans. Knowl. Data Eng. **30**(8), 1507–1518 (2018)
29. Grover, A., Leskovec, J.: node2vec: scalable feature learning for networks. In: Proceedings of the 22nd ACM SIGKDD International Conference on Knowledge Discovery and Data Mining, pp. 855–864. ACM, New York (2016)
30. Dey, A., Broumi, S., Bakali, A., Talea, M., Smarandache, F., et al.: A new algorithm for finding minimum spanning trees with undirected neutrosophic graphs. Granular Comput. **4**(1), 63–69 (2019)
31. Najman, L., Cousty, J., Perret, B.: Playing with kruskal: algorithms for morphological trees in edge-weighted graphs. In: International Symposium on Mathematical Morphology and Its Applications to Signal and Image Processing, pp. 135–146. Springer, Berlin (2013)
32. Debnath, A.K.., de Compadre, R.L.L., Debnath, R.L.L., Shusterman, A.J., Hansch, C.: Structure-activity relationship of mutagenic aromatic and heteroaromatic nitro compounds. Correlation with molecular orbital energies and hydrophobicity. J. Med. Chem. **34**(2), 786–797 (1991)
33. Toivonen, H., Srinivasan, A., King, R.D., Kramer, S., Helma, C.: Statistical evaluation of the predictive toxicology challenge 2000–2001. Bioinformatics **19**(10), 1183–1193 (2003)
34. Nguyen, D., Luo, W., Nguyen, T.D., Venkatesh, S., Phung, D.: Learning graph representation via frequent subgraphs. In: Proceedings of the 2018 SIAM International Conference on Data Mining, pp. 306–314. SIAM, Philadelphia (2018)
35. Sabour, S., Frosst, N., Hinton, G.: Matrix capsules with EM routing. In: 6th International Conference on Learning Representations, ICLR, pp. 1–15 (2018)

Chapter 4
Subgraph Augmentation with Application to Graph Mining

Jiajun Zhou, Jie Shen, Yalu Shan, Qi Xuan, and Guanrong Chen

Abstract Graph classification, which aims to identify the category labels of graphs, plays a significant role in drug classification, toxicity detection, protein analysis etc. However, the limitation of the general scale of benchmark datasets makes easily causes graph classification models to fall into overfitting and undergeneralization. In this chapter, the *M-Evolve* framework is introduced for graph classification (Zhou et al., Data augmentation for graph classification. In: Proceedings of the 29th ACM International Conference on Information & Knowledge Management, pp. 2341–2344, 2020; Zhou et al., M-evolve: structural-mapping-based data augmentation for graph classification. In: IEEE Transactions on Network Science and Engineering, pp. 1–1, 2020), a novel technique for expanding graph structured data spaces and optimizing graph classifiers. Typical graph tasks such as node classification and link prediction are unified to generate graph classification patterns, demonstrating some applications to multiple tasks in graph mining. One of the main contributions of this chapter is to apply the technique of subgraph augmentation for various tasks. The *M-Evolve* is general and flexible, which can be easily combined with existing graph classification models. Extensive experiments are conducted on real datasets to illustrate the effectiveness of our framework.

J. Zhou · J. Shen · Y. Shan · Q. Xuan (✉)
Institute of Cyberspace Security, Zhejiang University of Technology, Hangzhou, China

College of Information Engineering, Zhejiang University of Technology, Hangzhou, China
e-mail: xuanqi@zjut.edu.cn

G. Chen
Department of Electrical Engineering, City University of Hong Kong, Hong Kong SAR, China

4.1 Introduction

Graph classification, or network classification, has recently attracted considerable attention from different scientific and engineering communities such as bioinformatics [1] and chemoinformatics [2]. In bioinformatics, proteins or enzymes can be represented as labeled graphs, in which nodes are atoms and edges represent chemical bonds that connect atoms. The task of graph classification is to classify these molecular graphs according to their chemical properties like carcinogenicity, mutagenicity and toxicity.

However, in bioinformatics and chemoinformatics, the scale of the known benchmark graph datasets is generally in the range of tens to thousands, which is far from the scale of real-world social network datasets like COLLAB and IMDB [3]. Despite the advances of various graph classification methods, from graph kernels, graph embedding to graph neural networks, the limitation of data scale makes them easily fall into overfitting and undergeneralization. Overfitting refers to a modeling error that occurs when a model learns a function with high variance to perfectly fit the limited data. A natural idea to address overfitting is data augmentation, which is widely applied in computer vision (CV) and natural language processing (NLP). Data augmentation encompasses a number of techniques that enhance both the scale and the quality of training data such that the models of higher performance can be learnt satisfactorily. In CV, image augmentation methods include geometric transformation, color depth adjustment, neural style transfer, adversarial training, etc. However, different from image data, which have a clear grid structure, graphs have irregular topological structures, making it hard to generalize some basic augmentation operations to graphs.

To solve the above problem, we take an effective approach to study data augmentation on graphs and develop a subgraph augmentation method, called *motif-similarity mapping*. The idea is to generate more virtual data for small datasets via heuristic modification of graph structures. Since the generated graphs are artificial and treated as weakly labeled data, their validity remains to be verified. Therefore, we introduce a concept of label reliability, which reflects the matching degree between examples and their labels, to filter fine augmented examples from generated data. Furthermore, we introduce a model evolution framework, named *M-Evolve* [4, 5], which combines subgraph augmentation, data filtration and model retraining to optimize classifiers. We demonstrate that *M-Evolve* achieves a significant improvement of performance on graph classification.

We further explore the transferability of *M-Evolve* to other graph tasks such as node classification and link prediction. Motivated by recent works of graph neural networks (GNNs) [6, 7], we show that GNNs achieve node classification/link prediction by extracting a local subgraph around each target node/link, and by learning a function mapping the subgraph patterns to node labels/link existence. Hence, we unify both node classification and link prediction into graph classification, i.e., we achieve node classification or link prediction via local subgraph classification. We demonstrate that *M-Evolve* can be transferred to these graph tasks and achieve significant improvement of performance.

The rest of this chapter is organized as follows. In Sect. 4.2, we briefly review some related work on graph classification and data augmentation in graph mining. In Sect. 4.3, we introduce the subgraph augmentation technique and the model evolution framework. In Sect. 4.4, we discuss the application of the new framework to graph classification, node classification and link prediction, which involve a comparison between the results obtained from the original datasets and those obtained from subgraph augmentation. We conclude the chapter in Sect. 4.5.

4.2 Related Work

4.2.1 Graph Classification

4.2.1.1 Graph Kernel Methods

Graph kernels perform graph comparison by recursively decomposing pairwise graphs from the dataset into atomic substructures and then using a similarity function among these substructures. Intuitively, graph kernels can be understood as functions measuring the similarity of pairwise graphs. Generally, the kernels can be designed by considering various structural properties like the similarity of local neighborhood structures (WL kernel [8], propagation kernel [9]), the occurrence of certain graphlets or subgraphs (graphlet kernel [10]), the number of walks in common (random walk kernel [11–14]), and the attributes and lengths of the shortest paths (SP kernel [15]).

4.2.1.2 Embedding Methods

Graph embedding methods [16–19] capture the graph topology and derive a fixed number of features, ultimately achieving vector representations for the whole graph. For prediction on the graph level, this approach is compatible with any standard machine learning classifier such as support vector machine (SVM), k-nearest neighbors and random forest. Widely used embedding methods include graph2vec [20], structure2vec [21], subgraph2vec [22], etc.

4.2.1.3 Deep Learning Methods

Recently, increasing attention is drawn to the application of deep learning to graph mining and a wide variety of graph neural network (GNN) frameworks have been proposed for graph classification, including methods inspired by convolutional neural network (CNN), recurrent neural network (RNN), etc. One typical approach is to obtain a representation of the entire graph by aggregating the node embeddings that are the output of GNNs [2, 23]. Some sequential methods [24–26] handle these graphs with varying sizes by transforming them into sequences of fixed-length vectors and then feeding them to RNN. In addition, some hierarchical clustering methods [27–29] are used to learn hierarchical graph representations by combining GNNs with clustering algorithms. Notably, some recent works design universal graph pooling modules, which can learn the hierarchical representations of graphs and are compatible with various GNN architectures, e.g., DiffPool [30] learns a differentiable soft cluster assignment for vertices and then maps them to a coarsened graph layer by layer, and EigenPool [31] compresses the node features and local structures into coarsened signals via graph Fourier transform.

4.2.2 Data Augmentation in Graph Learning

Data augmentation, which aims to improve generalization of machine learning models, has been broadly studied in different fields such as computer vision (CV) and natural language processing (NLP). However, data augmentation for graph data remains under-explored. This is mainly due to the complex non-Euclidean structure of graphs, which limits possible manipulation operations. Recently, Zhao et al. [32] proposed a graph data augmentation framework named GAUG, which improves performance in GNN-based node classification via link prediction, based on these techniques that link prediction can effectively encode class-homophilic structure to promote intra-class edges and demote inter-class edges in given graph structure. Wang et al. [33] presented a node-parallel augmentation (NodeAug) strategy, which conducts data augmentation by changing both the node attributes and the graph structure, to improve the performance of GCN on semi-supervised node classification. Spinelli et al. [34] proposed a data augmentation strategy based on the application of GINN [35], in which a similarity graph is constructed using both labeled and unlabeled data and then a customized graph autoencoder is applied to augment new data. These studies focus on semi-supervised node classification task. While the work to be presented in this chapter is similar in motivation, it mainly tackles augmentation in the graph classification task.

4.3 The Model Evolution Framework for Graph Classification

4.3.1 Problem Formulation

Let $G = (V, E)$ be an undirected and unweighted graph, which consists of a node set $V = \{v_i \mid i = 1, \ldots, n\}$ and an edge set $E = \{e_i \mid i = 1, \ldots, m\}$. The topological structure of graph G is represented by an $n \times n$ adjacency matrix A with $A_{ij} = 1$ if $(i, j) \in E$ and $A_{ij} = 0$ otherwise. Dataset that contains a series of graphs

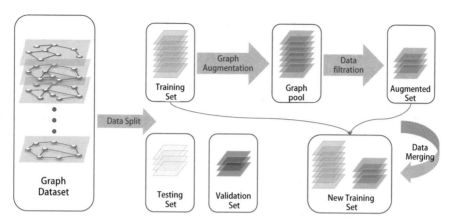

Fig. 4.1 An illustration of data augmentation application in graph (Subgraph Augmentation)

is denoted as $D = \{(G_i, y_i) \mid i = 1, \ldots, t\}$, where y_i is the label of graph G_i. For D, an upfront split will be applied to yield disjoint training, validation and testing set, denoted as D_{train}, D_{val} and D_{test}, respectively. The original classifier C will be pre-trained on D_{train} and D_{val}.

We further explore data augmentation technique for graph classification from a heuristic approach and consider optimizing graph classifier. Figure 4.1 demonstrates the application of data augmentation to graph structured data, which consists of two phases: subgraph augmentation and data filtration. Specifically, we aim to update a classifier with augmented data, which are first generated via subgraph augmentation and then filtered in terms of their label reliability. During subgraph augmentation, our purpose is to map the graph $G \in D_{train}$ to a new graph G' with the formal format: $f : (G, y) \mapsto (G', y)$. We treat the generated graphs as weakly labeled data and classify them into two groups via a label reliability threshold θ learnt from D_{val}. Then we merge the augmented set D'_{train}, filtered from generated graph pool D_{pool}, with D_{train} to produce the training set:

$$D_{train}^{new} = D_{train} + D'_{train} , \quad D'_{train} \subset D_{pool} . \tag{4.1}$$

Finally, we finetune or retrain the classifier with D_{train}^{new}, and evaluate it on the testing set D_{test}.

4.3.2 Subgraph Augmentation

Subgraph augmentation aims to expand training data via artificially creating more reasonable virtual data from a limited set of graphs. In this chapter, we consider augmentation as a topological mapping, which is conducted via heuristic modification of the graph structure. In order to ensure the approximate reasonability of the generated virtual data, our subgraph augmentation method will follow two principles: (1) edge modification, where G' is a partially modified graph with some of the edges added/removed from G; (2) structural property preservation, where augmentation operation keeps the graph connectivity and the number of edges constant. During edge modification, those edges removed from the graph are sampled from the candidate edge set E_{del}^c, while the edges added to the graph are sampled from the candidate pairwise nodes set E_{add}^c. The construction of candidate sets varies for different methods, as further discussed below.

4.3.2.1 Random Mapping

Here, consider *random mapping* as a simple baseline. The candidate sets are constructed as follows:

$$E_{del}^c = E , \quad E_{add}^c = \{(v_i, v_j) \mid A_{ij} = 0; \ i \neq j\}. \tag{4.2}$$

Notably, in a random scenario, E_{del}^c is the edge set of graph, and E_{add}^c is the set of virtual edges, which consist of unlinked pairwise nodes. Then, one can get the set of edges added/removed from G via sampling from the candidate sets randomly:

$$E_{del} = \{e_i \mid i = 1, \ldots, \lceil m \cdot \beta \rceil\} \subset E_{del}^c ,$$
$$E_{add} = \{e_i \mid i = 1, \ldots, \lceil m \cdot \beta \rceil\} \subset E_{add}^c , \tag{4.3}$$

where β is the budget of edge modification and $\lceil x \rceil = \mathbf{ceil}(x)$. Finally, based on the *random mapping*, the connectivity structure of the original graph is modified to generate a new graph:

$$G' = (V, (E \cup E_{add}) \backslash E_{del}) . \tag{4.4}$$

4.3.2.2 Motif-Similarity Mapping

Graph motifs are subgraphs that repeat themselves in a specific graph or even among various graphs. Each of these sub-graphs, defined by a particular pattern of interactions between nodes, may describe a framework, in which particular functions are achieved efficiently. Here, for simplicity, only consider open-triad motifs with chain structures. As shown on the left of Fig. 4.2, open-triad $\wedge_{ij}^a$ is equivalent to length-2 paths emanating from the head node v_i that induce a triangle.

The *motif-similarity mapping* aims to finetune these motifs to be approximately equivalent ones via edge swapping. During edge swapping, edge addition takes effect between the head and the tail nodes of the motif, while edge deletion removes an edge in the motif via weighted random sampling. For all open-triad motifs $\wedge_{ij}$, which has head node v_i and tail node v_j, the candidate set of pairwise nodes is

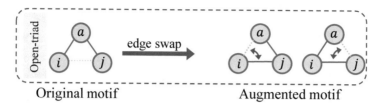

Fig. 4.2 Open-triad motif and heuristic edge swapping

denoted as

$$E_{add}^c = \{(v_i, v_j) \mid A_{ij} = 0, A_{ij}^2 \neq 0; i \neq j\}. \tag{4.5}$$

Then, we get E_{add}, the set of edges added to G, via weighted random sampling from E_{add}^c. For each $\wedge_{ij}$ involving pairwise nodes (v_i, v_j) in E_{add}, we remove one edge from it via weighted random sampling, and all of these removed edges constitute E_{del}.

Note that we assign all entries in E_{add}^c and $\wedge_{ij}$ with relative sampling weights that are associated with the node similarity scores. Specifically, before sampling, we compute the similarity scores over all entries in E_{add}^c using the *Resource Allocation* *(RA)* index, which has superiority among several local similarity indices [36]. For each entry (v_i, v_j) in E_{add}^c, the *RA* score s_{ij} and addition weight w_{ij}^{add} can be computed as follows:

$$
\begin{aligned}
s_{ij} &= \sum_{z \in \Gamma(i) \cap \Gamma(j)} \frac{1}{d_z}, \\
S &= \{s_{ij} \mid \forall (v_i, v_j) \in E_{add}^c\}, \\
w_{ij}^{add} &= \frac{s_{ij}}{\sum_{s \in S} s}, \\
W_{add} &= \{w_{ij}^{add} \mid \forall (v_i, v_j) \in E_{add}^c\},
\end{aligned}
\tag{4.6}
$$

where $\Gamma(i)$ denotes the one-hop neighbors of v_i and d_z denotes the degree of node z. Weighted random sampling means that the probability for an entry in E_{add}^c to be selected is proportional to its addition weight w_{ij}^{add}. Similarly, during edge deletion, the probability of edge sampled from $\wedge_{ij}$ is proportional to the deletion weight w_{ij}^{del}, as follows:

$$
\begin{aligned}
w_{ij}^{del} &= 1 - \frac{s_{ij}}{\sum_{s \in S} s}, \\
W_{del} &= \{w_{ij}^{del} \mid \forall (v_i, v_j) \in \wedge_{ij}\},
\end{aligned}
\tag{4.7}
$$

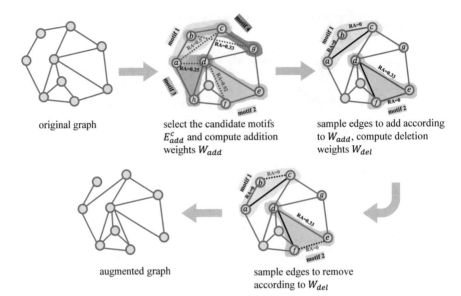

Fig. 4.3 An example of subgraph augmentation via motif-similarity mapping; red lines is the candidates and black lines is the modified edges

which means that edges with smaller *RA* scores have more chance to be removed. It is worth noting that many other similarity indices such as *Common Neighbors* (CN) and *Katz* [36] can also be applied into this scheme. Finally, the augmented graph can be obtained via Eq. (4.4). The example of subgraph augmentation using motif-similarity mapping is shown in Fig. 4.3.

Algorithm 1: Motif-similarity mapping

Input: Target network G, length of motif l, proportion of modification β.
Output: Augmented graph G'

1 Get E^c_{add} via Eq. (4.5) ;
2 Compute the addition weights W_{add} via Eq. (4.6) ;
3 $E_{add} \leftarrow$ weightRandomSample(E^c_{add}, $\lceil m \cdot \beta \rceil$, W_{add}) ;
4 Initialize $E^c_{del} = \emptyset$;
5 for $each\,(v_i, v_j) \in E_{add}$ **do**
6 $\quad$ Get the length-l motif h^l_{ij} via path search: $h^l_{ij} \leftarrow$ pathSearch(i, j, l) ;
7 $\quad$ Compute the deletion weights W_{del} via Eq. (4.7) ;
8 $\quad$ $e_{del} \leftarrow$ weightRandomSample(h^l_{ij}, 1, W_{del}) ;
9 $\quad$ Add e_{del} to E^c_{del} ;
10 end
11 Get augmented graph G' via Eq. (4.4) ;
12 end ;
13 return G';

4.3.3 Data Filtration

Due to the topological dependency of graph structured data, the examples generated via subgraph augmentation may lose some original semantics. By assigning the label of the original graph to the generated graph during subgraph augmentation, one cannot determine whether the assigned label is reliable. Therefore, the concept of label reliability is employed here to measure the matching degree between examples and labels.

Each graph G_i in D_{val} will be fed into classifier C to obtain the prediction vector $\mathbf{p}_i \in \mathbb{R}^{|Y|}$, which represents the probability distribution as how likely an input example belongs to each possible class. Here, $|Y|$ is the number of classes for labels. Then, a probability confusion matrix $\mathbf{Q} \in \mathbb{R}^{|Y| \times |Y|}$, in which the entry q_{ij} represents the average probability that the classifier classifies the graphs of the i-th class to the j-th class, is computed as follows:

$$\mathbf{q}_k = \frac{1}{\Omega_k} \sum_{y_i=k} \mathbf{p}_i ,$$

$$\mathbf{Q} = [\mathbf{q}_1, \mathbf{q}_2, \ldots, \mathbf{q}_{|Y|}] ,$$

(4.8)

where Ω_k is the number of graphs belonging to the k-th class in D_{val} and $\mathbf{q}_k$ is the average probability distribution of the k-th class.

The label reliability of an example (G_i, y_i) is defined as the product of example probability distribution $\mathbf{p}_i$ and class probability distribution $\mathbf{q}_{y_i}$ as follows:

$$r_i = \mathbf{p}_i^\top \mathbf{q}_{y_i} .$$

(4.9)

A threshold θ used to filter the generated data is defined as

$$\theta = \arg\min_{\theta} \sum_{(G_i, y_i) \in D_{val}} \Phi[(\theta - r_i) \cdot g(G_i, y_i)] ,$$

(4.10)

where $g(G_i, y_i) = 1$ if $C(G_i) = y_i$ and $g(G_i, y_i) = -1$ otherwise, and $\Phi(x) = 1$ if $x > 0$ and $\Phi(x) = 0$ otherwise.

4.3.4 Model Evolution Framework

Model evolution aims to iteratively optimize classifiers via subgraph augmentation, data filtration and model retraining, and ultimately improve the performance of graph classification. Figure 4.4 and Algorithm 2 demonstrate the workflow and the procedure of M-Evolve, respectively. Here, a variable, number of iterations T, is introduced for repeating the above workflow to continuously augment the dataset and optimize the classifier.

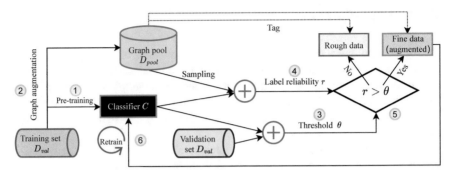

Fig. 4.4 The architecture of the model evolution. The complete workflow proceeds as follows: (1) pre-train graph classifier using training set; (2) apply subgraph augmentation to generate data pool; (3) compute the label reliability threshold using validation set; (4) compute the label reliability of examples sampled from graph pool; (5) filter data and obtain augmented set using threshold; (6) retrain graph classifier using the union of training set and augmented set

---

Algorithm 2: M-Evolve

Input: Training set D_{train}, validation set D_{val}, subgraph augmentation f, number of iterations T.

Output: Evolutive model C'

1   Pre-training classifier C using D_{train} and D_{val} ;

2   Initalize $iteration = 0$;

3   **for** $iteration < T$ **do**

4   Graph augmentation: $D_{pool} \leftarrow f(D_{train})$;

5   For all graphs G_i in D_{val} classified by C, get $\mathbf{p}_i$;

6   Get probability confusion matrix $\mathbf{Q}$ via Eq. (4.8) ;

7   For all graphs G_i in D_{val} classified by C, get r_i via Eq. (4.9);

8   Get the label reliability threshold θ via Eq. (4.10) ;

9   For all samples (G_i, y_i) in D_{pool} classified by C, compute r_i, **if** $r_i > \theta$, $D_{train}.append((G_i, y_i))$;

10   Get evolutive classifier: $C' \leftarrow$ retrain(C, D_{train}) ;

11      $iteration \leftarrow iteration + 1$;

12      $C \leftarrow C'$;

13 **end**

14 **end** ;

15 **return** C';

4.4   Application of Subgraph Augmentation

The *M-Evolve* framework is compatible with different graph classification models. In order to explore the transferability of *M-Evolve* to other graph tasks, we unify both node classification and link prediction into graph classification pattern, as shown in Fig. 4.5. For node classification/link prediction, we first extract the local subgraph around each target node/link, and all the extracted subgraphs are treated as dataset to train a graph classifier, which learns a function mapping the subgraph patterns to node labels/link existence. In this chapter, we choose a recent deep neural architecture DGCNN [37] as the default graph classifier.

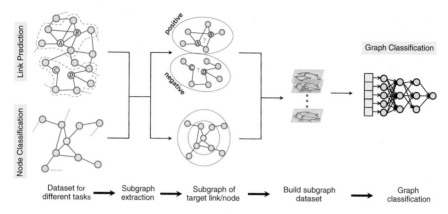

Fig. 4.5 Unify multiple tasks into graph classification. For link prediction, extract the local subgraph of target pairwise nodes, and the labels of subgraphs reflect the link existence. For node classification, extract the local subgraph of target nodes. The subgraph labels are equivalent to the corresponding node labels

4.4.1 Graph Classification

Graph classification focuses on identifying the category labels of graphs in a dataset, and plays a significant role in drug classification, toxicity detection, protein analysis etc. Widely used methods concentrate on graph kernels [8, 9, 14], graph embedding [16–18] and graph pooling [30, 31].

4.4.1.1 Experimental Setting

Data We evaluate the proposed methods on two benchmark datasets: Mutag [38] and PTC-MR [39]. The two datasets represent the graph collections of chemical compounds, in which nodes correspond to molecular structures and edges indicate chemical bonds between them. The specifications of datasets are given in Table 4.1.

Table 4.1 Dataset properties, where $|D|$ is the number of graphs in dataset, $|Y|$ is the number of classes for labels, $Avg.|V| / Avg.|E|$ is the average number of nodes/edges, *bias* is the proportion of the dominant class and *Attri* is the feature dim of the node in dataset

Task	Dataset	$|D|$	$|Y|$	Avg. $|V|$	Avg. $|E|$	bias (%)	Attri
Graph classification	MUTAG	188	2	17.93	19.79	66.5	–
	PTC-MR	344	2	14.29	14.69	55.8	–
Link prediction	Router	3822	2	50.00	50.15	50	–
	Celegans	2964	2	90.11	225.25	50	–
Node classification	BlogCatalog	5196	6	67.11	359.57	16.7	8189
	Flickr	7575	9	64.30	1139.25	11.1	12,047

Parameter Settings We first split each dataset into training, validation and testing sets with a proportion of 7:1:2. We re-split the experimental dataset 25 times and report the average accuracy across all trials. We use the default setting of DGCNN architecture, i,e., four graph convolution layers with 32,32,32,1 channels and a dense layer with 128 neurons.

4.4.2 Link Prediction

Link prediction, which aims to uncover missing links or predicting future inter-actions between pairwise nodes based on observable links and other external information, has been applied to friend recommendation [40], knowledge graph completion [41] and network reconstruction [42]. Widely used methods concentrate on similarity-based algorithms [36], maximum likelihood methods [43, 44], probabilistic models [45, 46] and graph autoencoder (GAE) [47, 48].

4.4.2.1 Subgraph Extraction

For an original graph $G = (V, E)$, given target pairwise nodes $v_i, v_j \in V$, the h-hop subgraph for (v_i, v_j) is the $G_{i,j}^h$ induced from G by the set of nodes $\{v_k \mid d(v_k, v_i) \leq h \text{ or } d(v_k, v_j) \leq h\}$, where $d(a, b)$ is the length of shortest path between nodes a and b. The subgraph $G_{i,j}^h$ describes the "h-hop surrounding environment" of the target link (v_i, v_j).

Node Importance Labeling Since the size of a subgraph is constrained by the radiation order of the target link, nodes with different relative positions to the target link have different structural importance, i.e., having different contributions to the prediction of link existence. Empirically, in a graph, pairwise nodes that are close to each other generally have higher influence than those that are far apart, so the natural to set the node importance label based on the distance of nodes from the central pairwise nodes (target link) in the subgraph. We set the node importance label using the Double-Radius Node Labeling (DRNL) criterion [7]:

$$f_{lp}(k) = 1 + \min(d_i, d_j) + (d/2)[(d/2) + (d\%2) - 1] \tag{4.11}$$

where $d_i := d(k, i)$, $d_j := d(k, j)$, $d := d_i + d_j$, $(d/2)$ and $(d\%2)$ are the integer quotient and remainder of d divided by 2. And for those nodes with $d(k, i) = \pm\infty$ or $d(k, j) = \pm\infty$, we give them a null label 0. Figure 4.6 shows an example of node importance labeling via DRNL. These node importance labels are uniquely hot-coded and followed by the node's own attribute features as node features for subsequent graph classification tasks.

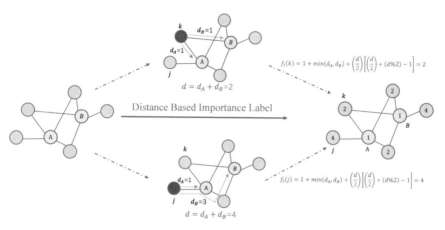

Fig. 4.6 An example for node importance labeling in link prediction

4.4.2.2 Experimental Setting

Data We evaluate the proposed methods on two benchmark datasets: Router [49] and Celegans [50]. Router is router-level Internet and Celegans is the neural network of Celegans. The specifications of datasets are given in Table 4.1.

Baselines To verify the effectiveness of DGCNN in solving link prediction problem, we compare DGCNN with GNN-based link prediction methods: graph autoencoder (GAE) and variational graph autoencoder (VGAE) [47].

Parameter Settings We split each dataset into training, validation and testing sets with a proportion of 1:1:3. We re-split the experimental dataset 25 times and report the average accuracy across all trials. We adjust the setting of DGCNN architecture, i,e., four graph convolution layers with 32,32,32,1 channels and a dense layer with 256 neurons. For subgraph extraction, we set the neighborhood hop counts h to 2.

4.4.3 Node Classification

Node classification is typically a semi-supervised learning task, in which only a few nodes' labels are known for predicting the labels of other nodes in graph datasets. It is widely used for recommendation, entity alignment of knowledge graphs, etc. Common methods include walk-based embedding methods [51, 52], spectral domain convolution methods [53], and spatial domain convolution methods [54].

4.4.3.1 Subgraph Extraction

For an original graph $G = (V, E)$, given a target node $v_i \in V$, the h-hop subgraph for v_i is the G_i^h induced from G by the set of nodes $\{v_k \mid d(v_k, v_i) \leq h\}$.

Node Importance Labeling Similar to the node importance labeling in link prediction task, we set the importance label of nodes by their distances from the target node:

$$f_{nc}(k) = d(k, i) . \tag{4.12}$$

4.4.3.2 Experimental Setting

Data We evaluate the proposed methods on two social datasets: BlogCatalog and flickr [55], in which the posted keywords or tags are used as node attribute information. The specifications of datasets are given in Table 4.1.

Baseline To verify the effectiveness of DGCNN in solving link prediction problem, we compare DGCNN with popular node classification deep models: graph convolution network (GCN) [6] and graph attention network (GAT) [56].

Parameter Settings We also split each dataset into training, validation and testing sets with a proportion of 1:2:7, We re-split the experimental dataset 25 times and report the average accuracy across all trials. We adjust the setting of DGCNN architecture, i,e., two graph convolution layers with 32,1 channels and a dense layer with 256 neurons. For subgraph extraction, we set the neighborhood hop counts h to 1.

4.4.4 Experimental Results

We unify multiple tasks into graph classification pattern, and use DGCNN architecture to complete all tasks by graph classification. Tables 4.2, 4.3 and 4.4 report the results of performance comparison among baselines, DGCNN and the evolutive models for different tasks.

First, from the comparison between DGCNN and the evolutive models, one can see that there is a boost in classification performance across all six datasets, indicating that our *M-Evolve* can effectively improve the performance of graph classification model. We speculate that the original DGCNN models trained with limited training data are overfitting, and on the contrary, *M-Evolve* enriches the scale of training data via subgraph augmentation and optimizes graph classifiers via iterative retraining, therefore can improve the generalization and avoid overfitting to a certain extent.

Table 4.2 Graph classification results of original and evolutive models. The best results are marked in bold. The far-right column gives the average relative improvement rate (Avg. RIMP) in accuracy

Dataset	Mapping	Budget			Avg. RIMP
		0.10	**0.15**	**0.20**	
MUTAG	DGCNN	0.8447			–
	DGCNN + random	0.8447	0.8533	**0.8458**	+0.38%
	DGCNN + m-s	**0.8450**	**0.8547**	0.8436	+0.36%
PTC_MR	DGCNN	0.5775			–
	DGCNN + random	0.5739	0.5764	**0.5860**	+0.22%
	DGCNN + m-s	**0.5849**	**0.5962**	0.5733	+1.26%

Table 4.3 Link prediction results in baselines, DGCNN and evolutive models. The best results are marked in bold. The far-right column gives the average relative improvement rate (Avg. RIMP) in accuracy

Dataset	Mapping	Budget			Avg. RIMP
		0.10	**0.15**	**0.20**	
Router	GAE	0.5130			–
	VGAE	0.4999			–
	DGCNN	0.6721			–
	DGCNN + random	0.6430	0.6512	0.6694	−2.6%
	DGCNN + m-s	**0.6858**	**0.6852**	**0.6854**	+1.7%
Celegans	GAE	0.5256			–
	VGAE	0.5053			–
	DGCNN	0.6323			–
	DGCNN + random	0.6170	0.6125	0.6176	−2.6%
	DGCNN + m-s	**0.6353**	**0.6379**	**0.6379**	+0.7%

Table 4.4 Node classification results in baselines, DGCNN and evolutive models. The best results are marked in bold. The far-right column gives the average relative improvement rate (Avg. RIMP) in accuracy

Dataset	Mapping	Budget			Avg. RIMP
		0.10	**0.15**	**0.20**	
Blog	GCN	0.7200			–
	GAT	0.6630			–
	DGCNN	0.7453			–
	DGCNN + random	0.7502	0.7493	**0.7483**	+0.53%
	DGCNN + m-s	**0.7589**	**0.7560**	0.7457	+1.10%
Flickr	GCN	**0.5460**			–
	GAT	0.3590			–
	DGCNN	0.4192			–
	DGCNN + random	0.4471	0.4499	0.4505	+7.15%
	DGCNN + m-s	0.4888	0.4884	0.5014	+17.57%

Now, we define the relative improvement rate (RIMP) in accuracy as follows:

$$RIMP = \frac{Acc_{en} - Acc_{ori}}{Acc_{ori}}, \tag{4.13}$$

where Acc_{en} and Acc_{ori} refer to the accuracy of the evolutive and the original models, respectively. In Tables 4.2, 4.3 and 4.4, the far-right column gives the average relative improvement rate (Avg. *RIMP*) in accuracy, from which one can see that the *M-Evolve* combined with motif-similarity (m-s) mappings obtains better results overall. These results indicate that both similarity and motif mechanisms play positive roles in enhancing graph classification. As a reasonable explanation, similarity mechanism tends to link nodes with higher similarity and is capable of optimizing topological structure legitimately, which is similar to the finding method in [57]. And the motif mechanism achieves edge modification via local edge swapping, which has subtle effect on both the degree distribution and the clustering coefficient of the graph.

We further investigate the effectiveness of unifying both link prediction and node classification tasks into graph classification patterns. Specifically, we compare the performances of DGCNN and baselines. Notably, we use small training set that only accounts for 20%/10% of the entire dataset on link prediction/node classification. For link prediction, DGCNN significantly outperforms baselines since both GAE and VGAE have poor performances when being trained with extremely sparse graphs. For node classification, DGCNN outperforms baselines on Blog, but has poorer performance than GCN. As a reasonable explanation, when constructing the subgraph dataset, we extract 1-hop subgraph in order to reduce the running cost of the algorithm, while GCN aggregates the 2-hop neighborhoods and uses more neighborhood information. Overall, DGCNN achieves competitive performances against baselines in both link prediction and node classification tasks, indicating that the idea of unifying both link prediction and node classification tasks into graph classification is effective.

4.5   Conclusion

In this chapter, we unify node classification and link prediction tasks into graph classification patterns, and introduce a concept of subgraph augmentation to generate weakly labeled data for small benchmark datasets through heuristic transformations of graph structures. In addition, we propose a general model evolution framework that combines subgraph augmentation, data filtering, and model retraining, to optimize a pre-trained graph classifier. Experiments on six datasets show that our proposed framework performs well and helps existing graph classification models mitigate the overfitting problem when being trained on small-scale benchmark datasets, and achieve significant improvement in classification performance.

References

1. Borgwardt, K.M., Ong, C.S., Schönauer, S., Vishwanathan, S.V.N., Smola, A.J., Kriegel, H.-P.: Protein function prediction via graph kernels. Bioinformatics **21**(suppl_1), i47–i56 (2005)
2. Duvenaud, D.K., Maclaurin, D., Iparraguirre, J., Bombarell, R., Hirzel, T., Aspuru-Guzik, A., Adams, R.P.: Convolutional networks on graphs for learning molecular fingerprints. In: Advances in Neural Information Processing Systems, pp. 2224–2232 (2015)
3. Yanardag, P., Vishwanathan, S.V.N.: Deep graph kernels. In: Proceedings of the 21th ACM SIGKDD International Conference on Knowledge Discovery and Data Mining, pp. 1365–1374 (2015)
4. Zhou, J., Shen, J., Xuan, Q.: Data augmentation for graph classification. In: Proceedings of the 29th ACM International Conference on Information & Knowledge Management, pp. 2341–2344 (2020)
5. Zhou, J., Shen, J., Yu, S., Chen, G., Xuan, Q.: M-evolve: structural-mapping-based data augmentation for graph classification. In: IEEE Transactions on Network Science and Engineering, pp. 1–1 (2020)
6. Kipf, T.N., Welling, M.: Semi-supervised classification with graph convolutional networks. In: International Conference on Learning Representations (ICLR) (2017)
7. Zhang, M., Chen, Y.: Link prediction based on graph neural networks. In: Advances in Neural Information Processing Systems, pp. 5165–5175 (2018)
8. Shervashidze, N., Schweitzer, P., Van Leeuwen, E.J., Mehlhorn, K., Borgwardt, K.M.: Weisfeiler-Lehman graph kernels. J. Mach. Learn. Res. **12**(9) (2011)
9. Neumann, M., Garnett, R., Bauckhage, C., Kersting, K.: Propagation kernels: efficient graph kernels from propagated information. Mach. Learn. **102**(2), 209–245 (2016)
10. Shervashidze, N., Vishwanathan, S.V.N., Petri, T., Mehlhorn, K., Borgwardt, K.: Efficient graphlet kernels for large graph comparison. In: Artificial Intelligence and Statistics, pp. 488–495 (2009)
11. Gärtner, T., Flach, P., Wrobel, S.: On graph kernels: hardness results and efficient alternatives. In: Learning Theory and Kernel Machines, pp. 129–143 (Springer, Berlin, 2003)
12. Kashima, H., Tsuda, K., Inokuchi, A.: Marginalized kernels between labeled graphs. In: Proceedings of the 20th International Conference on Machine Learning (ICML-03), pp. 321–328 (2003)
13. Mahé, P., Ueda, N., Akutsu, T., Perret, J.-L., Vert, J.-P.: Extensions of marginalized graph kernels. In: Proceedings of the twenty-first International Conference on Machine Learning, p. 70 (2004)
14. Sugiyama, M., Borgwardt, K.: Halting in random walk kernels. In: Advances in Neural Information Processing Systems, pp. 1639–1647 (2015)
15. Borgwardt, K.M., Kriegel, H.-P.: Shortest-path kernels on graphs. In: Fifth IEEE International Conference on Data Mining (ICDM'05), p. 8. IEEE, Piscataway (2005)
16. Cai, H., Zheng, V.W., Chang, K.C.-C.: A comprehensive survey of graph embedding: problems, techniques, and applications. IEEE Trans. Knowl. Data Eng. **30**(9), 1616–1637 (2018)
17. Chen, F., Wang, Y.-C., Wang, B., Jay Kuo, C.-C.: Graph representation learning: a survey. In: APSIPA Transactions on Signal and Information Processing, vol. 9 (2020)
18. Fu, C., Zheng, Y., Liu, Y., Xuan, Q., Chen, G.: Nes-tl: network embedding similarity-based transfer learning. IEEE Trans. Netw. Sci. Eng. **7**(3), 1607–1618 (2019)
19. Guo, W., Shi, Y., Wang, S., Xiong, N.N.: An unsupervised embedding learning feature representation scheme for network big data analysis. IEEE Trans. Netw. Sci. Eng. **7**(1), 115–126 (2019)
20. Narayanan, A., Chandramohan, M., Venkatesan, R., Chen, L., Liu, Y., Jaiswal, S.: graph2vec: learning distributed representations of graphs. Preprint. arXiv:1707.05005 (2017)
21. Dai, H., Dai, B., Song, L.: Discriminative embeddings of latent variable models for structured data. In: International Conference on Machine Learning, pp. 2702–2711 (2016)

22. Narayanan, A., Chandramohan, M., Chen, L., Liu, Y., Saminathan, S.: subgraph2vec: learning distributed representations of rooted sub-graphs from large graphs. Preprint. arXiv:1606.08928 (2016)
23. Gilmer, J., Schoenholz, S.S., Riley, P.F., Vinyals, O., Dahl, G.E.: Neural message passing for quantum chemistry. Preprint. arXiv:1704.01212 (2017)
24. Li, Y., Tarlow, D., Brockschmidt, M., Zemel, R.: Gated graph sequence neural networks. Preprint. arXiv:1511.05493 (2015)
25. Jin, Y., JaJa, J.F.: Learning graph-level representations with recurrent neural networks. Preprint. arXiv:1805.07683 (2018)
26. You, J., Ying, R., Ren, X., Hamilton, W.L., Leskovec, J.: GraphRNN: generating realistic graphs with deep auto-regressive models. Preprint. arXiv:1802.08773 (2018)
27. Defferrard, M., Bresson, X., Vandergheynst, P.: Convolutional neural networks on graphs with fast localized spectral filtering. In: Advances in Neural Information Processing Systems, pp. 3844–3852 (2016)
28. Simonovsky, M., Komodakis, N.: Dynamic edge-conditioned filters in convolutional neural networks on graphs. In: Proceedings of the IEEE Conference on Computer Vision and Pattern Recognition, pp. 3693–3702 (2017)
29. Fey, M., Lenssen, J.E., Weichert, F., Müller, H.: SplineCNN: fast geometric deep learning with continuous b-spline kernels. In: Proceedings of the IEEE Conference on Computer Vision and Pattern Recognition, pp. 869–877 (2018)
30. Ying, Z., You, J., Morris, C., Ren, X., Hamilton, W., Leskovec, J.: Hierarchical graph representation learning with differentiable pooling. In: Advances in Neural Information Processing Systems, pp. 4800–4810 (2018)
31. Ma, Y., Wang, S., Aggarwal, C.C., Tang, J.: Graph convolutional networks with eigenpooling. In: Proceedings of the 25th ACM SIGKDD International Conference on Knowledge Discovery and Data Mining, pp. 723–731 (2019)
32. Zhao, T., Liu, Y., Neves, L., Woodford, O., Jiang, M., Shah, N.: Data augmentation for graph neural networks. Preprint. arXiv:2006.06830 (2020)
33. Wang, Y., Wang, W., Liang, Y., Cai, Y., Liu, J., Hooi, J.: NodeAug: semi-supervised node classification with data augmentation. In: Proceedings of the 26th ACM SIGKDD International Conference on Knowledge Discovery & Data Mining, pp. 207–217 (2020)
34. Spinelli, I., Scardapane, S., Scarpiniti, M., Uncini, A.: Efficient data augmentation using graph imputation neural networks. In: Progresses in Artificial Intelligence and Neural Systems, pp. 57–66 (Springer, Berlin, 2020)
35. Spinelli, I., Scardapane, S., Uncini, A.: Missing data imputation with adversarially-trained graph convolutional networks. Neural Netw. **129**, 249–260 (2020)
36. Zhou, T., Lü, L., Zhang, Y.-C.: Predicting missing links via local information. Eur. Phys. J. B **71**(4), 623–630 (2009)
37. Zhang, M., Cui, Z., Neumann, M., Chen, Y.: An end-to-end deep learning architecture for graph classification. In: Thirty-Second AAAI Conference on Artificial Intelligence (2018)
38. Debnath, A.K., Lopez de Compadre, R.L., Debnath, G., Shusterman, A.J., Hansch, C.: Structure-activity relationship of mutagenic aromatic and heteroaromatic nitro compounds. Correlation with molecular orbital energies and hydrophobicity. J. Med. Chem. **34**(2), 786–797 (1991)
39. Helma, C., King, R.D., Kramer, S., Srinivasan, A.: The predictive toxicology challenge 2000–2001. Bioinformatics **17**(1), 107–108 (2001)
40. Adamic, L.A., Adar, E.: Friends and neighbors on the web. Soc. Netw. **25**(3), 211–230 (2003)
41. Nickel, M., Murphy, K., Tresp, V., Gabrilovich, E.: A review of relational machine learning for knowledge graphs. Proc IEEE **104**(1), 11–33 (2015)
42. Oyetunde, T., Zhang, M., Chen, Y., Tang, Y., Lo, C.: BoostGAPFILL: improving the fidelity of metabolic network reconstructions through integrated constraint and pattern-based methods. Bioinformatics **33**(4), 608–611 (2017)
43. Clauset, A., Moore, C., Newman, M.E.J.: Hierarchical structure and the prediction of missing links in networks. Nature **453**(7191), 98–101 (2008)

44. White, H.C., Boorman, S.A., Breiger, R.L.: Social structure from multiple networks. I. Block-models of roles and positions. Am. J. Soc. **81**(4), 730–780 (1976)
45. Getoor, L., Friedman, N., Koller, D., Pfeffer, A.: Learning probabilistic relational models. In: Relational Data Mining, pp. 307–335. Springer, Berlin (2001)
46. Heckerman, D., Meek, C., Koller, D.: Probabilistic entity-relationship models, PRMS, and plate models. In: Introduction to Statistical Relational Learning, pp. 201–238 (2007)
47. Kipf, T.N., Welling, M.: Variational graph auto-encoders. Preprint. arXiv:1611.07308 (2016)
48. Pan, S., Hu, R., Long, G., Jiang, J., Yao, L., Zhang, C.: Adversarially regularized graph autoencoder for graph embedding. Preprint. arXiv:1802.04407 (2018)
49. Spring, N., Mahajan, R., Wetherall, D., Anderson, T.: Measuring ISP topologies with rocket-fuel. IEEE/ACM Trans. Netw. **12**(1), 2–16 (2004)
50. Watts, D.J., Strogatz, S.H.: Collective dynamics of 'small-world' networks. Nature **393**(6684), 440–442 (1998)
51. Perozzi, B., Al-Rfou, R., Skiena, S.: Deepwalk: online learning of social representations. In: Proceedings of the 20th ACM SIGKDD International Conference on Knowledge Discovery and Data Mining, pp. 701–710 (2014)
52. Grover, A., Leskovec, J.: node2vec: Scalable feature learning for networks. In: Proceedings of the 22nd ACM SIGKDD International Conference on Knowledge Discovery and Data Mining, pp. 855–864 (2016)
53. Kipf, T.N., Welling, M.: Semi-supervised classification with graph convolutional networks. Preprint. arXiv:1609.02907 (2016)
54. Hamilton, W., Ying, Z., Leskovec, J.: Inductive representation learning on large graphs. In: Advances in Neural Information Processing Systems, pp. 1024–1034 (2017)
55. Wu, J., He, J., Xu, J.: Demo-net: degree-specific graph neural networks for node and graph classification. In: Proceedings of the 25th ACM SIGKDD International Conference on Knowledge Discovery and Data Mining. ACM, New York (2019)
56. Veličković, P., Cucurull, G., Casanova, A., Romero, A., Liò, P., Bengio, Y.: Graph attention networks. In: International Conference on Learning Representations (2018). A poster
57. Zheng, M., Allard, A., Hagmann, P., Alemán-Gómez, Y., Serrano, M.Á.: Geometric renormal-ization unravels self-similarity of the multiscale human connectome. Proc. Natl. Acad. Sci. **117**(33), 20244–20253 (2020)

Chapter 5
Adversarial Attacks on Graphs: How to Hide Your Structural Information

Yalu Shan, Junhao Zhu, Yunyi Xie, Jinhuan Wang, Jiajun Zhou, Bo Zhou, and Qi Xuan

Abstract Deep learning has enjoyed the status of crown jewels in artificial intelligence, showing an impressive performance in various fields, especially in computer vision. However, most deep learning models are vulnerable and easy to be fooled by some slight disturbances in the input, which are called adversarial attacks. As the deep learning models are extended to graphs, adversarial attacks also threaten various graph data mining tasks, e.g., node classification, link prediction, community detection, and graph classification. One can modify the topology or features of graphs, such as manipulating a few edges or nodes, to downgrade the performance of graph algorithms. The vulnerability of these algorithms may largely hinder their applications and thus receives tremendous attention. In the current chapter, we overview the existing researches on graph adversarial attacks. In particular, we briefly summarize and classify the existing graph adversarial attack methods, e.g., heuristic, gradient and reinforcement learning, and then choose several classic adversarial attack methods on different graph tasks for detailed introduction. And finally, we also summarize the challenges in this area.

We begin with the background of the adversarial attack in Sect. 5.1 and then introduce some basic concepts of adversarial attacks in Sect. 5.2. After that, we discuss some cases of adversarial attack to node classification, link prediction, community detection, and graph classification in Sect. 5.3. Finally, we summarize the challenges and outline several future research directions in Sect. 5.4.

Y. Shan · J. Zhu · Y. Xie · J. Wang · J. Zhou · B. Zhou · Q. Xuan (✉)
Institute of Cyberspace Security, Zhejiang University of Technology, Hangzhou, China

College of Information Engineering, Zhejiang University of Technology, Hangzhou, China
e-mail: xuanqi@zjut.edu.cn

5.1 Background

The past few years have witnessed the growing prevalence of deep learning on applications such as image recognition, object detection, and natural language processing. Meanwhile, increasing attention has been paid to several security scenarios like autonomous driving, in which deep learning shows excellent performance. However, the deep learning models frequently used in recent years have been proved unstable and unreliable since they are vulnerable to perturbations. Szegedy et al. [1] first noticed the existence of adversarial examples in image classification, i.e., slight about deliberate perturbations can pollute an image and thereby reduce the classification performance. Such unreliability seriously limits the applicability of the deep learning models. Extensive studies on adversarial attack have been presented to verify the vulnerability of deep learning models. For example, Goodfellow et al. [2] proposed a gradient-based method (FGSM) to generate adversarial image samples, which can significantly degrade the performance of deep learning models. Szegedy et al. [1] proposed an untargeted attack algorithm called DeepFool. DeepFool misclassifies the model by projecting an image to the closest separating hyperplane according to the least distortion (in the sense of Euclidean distance). Carlini and Wagner et al. [3] proposed the Carlini and Wagner (C&W) attack strategy, which is an optimization-based method. C&W method takes advantage of the internal configurations of a targeted DNN for attack guidance and uses the L2 norm (e.g., Euclidean distance) to quantify the difference between adversarial examples and original examples.

Despite adversarial attack methods have made significant progress from the heuristic and gradient-based methods to optimization-based methods, researchers mainly focused on computer vision tasks and remained under-explored for graph learning. As a powerful representation, graphs can model diverse data in different areas, such as biology (protein interaction networks), chemistry (molecules), and society (social networks). Adversarial attacks on graph data mining algorithms also make sense in reality. Take the online social networks as an example. The fake account reduces its suspiciousness by following normal accounts and publishing daily content to avoid being detected and blocked. A malicious user may manipulate his profile or connect to targeted users on purpose to mislead the model. Similarly, adding fake comments to specific products can fool the recommender systems of a website.

Here, we briefly review the existing adversarial attack methods on graphs. Zügner et al. [4] first proposed an attack method on the graph data, named NETTACK, against the graph convolutional network (GCN) for node classification. They crafted the adversarial graphs by utilizing incremental computations. After that, Zügner et al. [5] designed a highly scalable adversarial attack algorithm named FASTTACK. FASTTACK successfully creates adversarial perturbations which are effective across models and datasets by exploiting the statistically significant patterns in adversarial perturbations of NETTACK.

Among the existing adversarial attack strategies, gradient-based attack methods have attracted the most attention due to their simplicity and outstanding performance. Zügner et al. [6] proposed the meta attack, which decreases the GCN performance by treating input data as a hyperparameter. Chen et al. [7] developed the Momentum Gradient Attack (MGA) strategy since the momentum gradient can not only stabilize the updating direction but also make the model jump out of poor local optimum. In order to solve the problem caused by the discreteness on graphs, Wu et al. [8] introduced integrated gradients, which can accurately reflect the effect of attack operation. On link-level task, Chen et al. [9] used the Iterative Gradient Attack (IGA) to mislead the GAE model. As for dynamic network link prediction, Chen et al. [10] developed Time-aware Gradient Attack (TGA) and crafted adversarial examples by exploiting the gradient information generated by a deep dynamic network. Tang et al. [11] proposed an attack strategy against Hierarchical Graph Pooling Neural Networks for graph classification. They designed a surrogate model which consists of convolutional and pooling operators and set the preserved nodes by the pooling operator as the attack targets. Li et al. [12] introduced the Simplified Gradient-based Attack (SGA), which served the Simplified Graph Convolution Network (SGC) as the surrogate model. This method solves the problem of the high complexity of time and space when gradient-based attack strategies are applied to the large scale graph. Ma et al. [13] proposed a black-box attack strategy, which is adapted from the white-box setup gradient-based, and then they corrected the attacker's loss with a misclassification rate. Finkelshtein et al. [14] developed a white-box gradient-based approach and a black-box approach relied on graph topology only, and then they demonstrated that the effect of a single node attack is similar to that of a multi-node attack.

In addition to gradient-based methods, the Genetic Algorithm (GA), as another optimization method, is also applied to adversarial attack. Chen et al. [15] used a (GA)-based Q-Attack to destroy the community structure of a graph. Yu et al. [16] considered attacking node classification and community detection and introduced a GA-based Euclidean Distance Attack (EDA) strategy for graph embedding. Besides, Dai et al. [17] and Ma et al. [18] used reinforcement learning to learn the attack strategy against node classification and graph classification, respectively. Fan et al. [19] exploited reinforcement learning to develop a black-box attack strategy against dynamic network link prediction. What's more, GAN is applied to adversarial attacks on graph data [20, 21]. Chen et al. [21] adjusted the traditional GAN and considered an adaptive graph adversarial attack (AGA-GAN), which contains a multi-strategy generator, similarity discriminator, and attack discriminator, and used subgraph to reduce the cost of attack. There are also other attack algorithms, such as heuristic algorithms [22, 23]. For example, Yu et al. [22] proposed three heuristic strategies by rewiring. Zhou et al. [23] focused on two broad classes of such approaches, one uses only local information about target links, and the other uses global network information.

5.2 Adversarial Attack

According to the existing studies, we first introduce a general definition of adversarial attack against graph data mining algorithms. Then, we classify these attack strategies from different perspectives.

5.2.1 Problem Definition

The adversarial attack tries to degrade graph data mining models by slightly modifying a graph's topology or features. In the following, we will summarize and give a general formulation for the adversarial attack on graphs.

Denote a graph as $G = (A, X)$, where A is the adjacency matrix and X is the feature matrix of nodes. Let f represent a learning model for a graph task and let y_i denote the label/ground truth of the graph or node. The attacker aims to maximize the attack loss $\mathcal{L}_{atk}$ by crafting an adversarial example $\hat{G} = (\hat{A}, \hat{X})$ with imperceptible perturbations in $\hat{A}$ or $\hat{X}$.

$$\underset{\hat{G} \in \Psi(G)}{maximize} \sum_{i \in T} \mathcal{L}_{atk}(f_{\theta^*}(\hat{G}^i), y_i)$$

$$s.t. \theta^* = \underset{\theta}{argmin} \sum_{j \in L} \mathcal{L}_{train}(f_\theta(G'^j), y_j), \tag{5.1}$$

where $\Psi(G)$ indicates the space of the perturbations on G, and G' can be G or $\hat{G}$, representing the original graph and the perturbed one, respectively. T represents the test set and L is the training set. In order to make the attack as imperceptible as possible, we set a fixed budget Δ for limiting the size of the perturbations in the adversarial samples:

$$\mathcal{Q}(\hat{G}^i, G^i) < \Delta$$

$$s.t. \hat{G}^i \in \Psi(G), \tag{5.2}$$

where $\mathcal{Q}(\cdot)$ denotes the likelihood ratio function, such as the number of common neighbors of given nodes, cosine similarity, Jaccard similarity, and so on.

5.2.2 Taxonomies of Attacks

We briefly introduce the main taxonomies of adversarial attacks on graph data and then classify them into different categories based on the attacker's capability, attacker's knowledge, attack strategy, and attack goals.

Attacker's Capability There are two settings for an attack, i.e., evasion attack and poisoning attack. The main difference between them is the attackers' capacity to insert adversarial perturbations.

- **Poisoning Attack:** Attacking happens when the learning model is trained. Attackers focus on the training phase of a learning model, aiming to make the trained model have malfunctioned.
- **Evasion Attack:** The model is fixed, which is trained with clean graphs. The attacker attacks during the test time. Moreover, they cannot change the model parameters and the structure.

Attacker's Knowledge To conduct attacks on the target system, attackers will possess specific knowledge about the target models and the datasets, which helps them achieve the adversarial goal. Accordingly, we can classify them into three settings, i.e., white-box attack, gray-box attack, and black-box attack.

- **White-box Attack:** The attacker has all the information about the target model, including the model architecture, the parameters, and the gradient information, i.e., the target model is transparent to the attacker. However, white-box attacks are challenging to apply in real scenarios since it is almost impossible for the attacker to obtain all the information about the model.
- **Gray-box Attack:** This case is more realistic since attackers have limited knowledge about the target model. For example, attackers can only get the labels of training data, which can be used to train a surrogate model for conducting an attack, rather than the parameters. Therefore, comparing with the white-box attack, gray-box attack is less effective but more realistic.
- **Black-box Attack:** Under this setting, attackers know nothing about the target model and only allow queries on limited samples. Therefore, it is easiest to conduct a black-box attacks which is however least effective comparing with the other two other kinds of attacks.

Attack Strategy For attacking a target model on graph data, attackers may have various choices on strategies. Most studies mainly focus on modifying the topology or change the node/edge features. Accordingly, the existing works can be categorized as topology attack, feature attack, and hybrid attack.

- **Topology Attack:** Attackers mainly focus on attacking algorithms by manipulating the graph structure, i.e., adding edges/nodes, deleting edges/nodes, rewiring edges. Typically, adding nodes/edges will increase the network's size, while deleting edges/nodes increases the risk of network disconnection.
- **Feature Attack:** Graph data mining algorithms, especially some GNNs, utilize the topology and exploit the node/edge features. Therefore, attackers can also conduct an attack by changing these node/edge features. Unlike graph structure, the node/edge features could be either binary or continuous. Thus, different operations can be performed on the features, e.g., flipping them for binary features or adding a small value for continuous features.

- **Hybrid Attack:** In most cases, attackers will conduct both of the above attack strategies at the same time to make the attack more effective.

Attack Goal Generally, there are two different scenarios according to the attackers' purpose: only decreasing the target node/edge prediction or degrading the model performance on the whole test set. Thus, we can divide the attack strategies into the following categories.

- **Targeted Attack:** Attackers mainly focus on evasion attacks since they aim to let the well-trained model misclassify a few target samples. According to the attacker's knowledge, we can further divide the target attack into: (1) direct attack, where attackers perform operations on the target node directly, and (2) influencer attack where attackers can only manipulate other nodes, e.g., neighbor nodes, to influence the target node prediction.
- **Untargeted Attack:** To degrade the model's performance in the whole test set, attackers prefer to poison the trained model by putting poisons into the training set. Obviously, compared to the targeted attack, the untargeted attack has a wider influence and thus could be more dangerous.

5.3 Attack Strategy

There are four typical tasks on graph data mining: node classification, link prediction, community detection and graph classification. Next, we will introduce attack strategies on these four tasks and give the details for one or two classical attack strategies.

5.3.1 Node Classification

Node classification is one of the most common tasks in graph data mining. Given a graph and a set of labeled nodes, the purpose of the task is to predict the label of the unlabeled nodes. Node classification is widely applied in different fields. For example, we can use node classification to classify the role of a protein in the biological network or predict the customer type of a user in e-commerce networks. However, the vulnerability of graph algorithms hinders the development of node classification. Zügner et al. [4] considered both the test phase and training phase and generated the adversarial graphs by performing perturbations on node features and graph structure. Dai et al. [17] developed two effective attack strategies for node classification and graph classification by utilizing reinforcement learning and genetic algorithm. However, they restricted their attacks on edge deletions only and did not evaluate the transferability. Moreover, Bojchevski et al. [24] performed the poisoning attack by exploiting perturbation theory, which is used to maximize the loss obtained after training DeepWalk. Most works [6, 8, 20, 25, 26] were concentrated on gradient-based strategies and crafted adversarial examples by adding/deleting edges according to the gradient information. Here, we mainly introduce NETTACK and Meta attack in detail.

5.3.1.1 NETTACK

Zügner et al. [4] proposed an efficient algorithm named NETTACK to generate adversarial samples with imperceptible perturbation. NETTACK, a sequential approach, uses the surrogate model to generate adversarial graphs. Relevant experiments showed that the attack method can significantly degrade the performance of models on node classification task.

NETTACK adopted the classical GCN model as the surrogate model. Traditionally, a GCN model with a single hidden layer can be represented as follow:

$$Z = f_\theta(A, X) = softmax(\hat{A}\sigma(\hat{A}XW^{(1)})W^{(2)}), \tag{5.3}$$

where $\hat{A} = \tilde{D}^{-\frac{1}{2}}\tilde{A}\tilde{D}^{-\frac{1}{2}}$, $\tilde{A} = A + I_N$ is the adjacency matrix of graph G after adding self-loops via the identity matrix I_N, $W^{(l)}$ is the trainable weight matrix of layer l, $\sigma(\cdot)$ is an activation function like ReLU. To simplify the model, Zügner et al. used linear activation function and the surrogate model could be represented as

$$Z' = softmax(\hat{A}\hat{A}XW^{(1)}W^{(2)}) = softmax(\hat{A}^2XW), \tag{5.4}$$

where $\hat{A}^2XW$ is the log-probabilities of the nodes, which can affect the classifier performance. To downgrade the classification accuracy of the target node, attackers should calculate the log-probability of the target node and craft the adversarial graphs that have a significant difference from the original one in the node log-probability. Accordingly, the attacker's loss is denoted as

$$\mathcal{L}_{atk}(A, X; W, v_0) = \max_{y \neq y_{old}} [\hat{A}^2XW]_{v_0,y} - [\hat{A}^2XW]_{v_0,y_{old}}, \tag{5.5}$$

where y is the adversarial label and y_{old} is the clean label, both of them are predicted by the surrogate model. The goal of attackers is to maximize the attacker's loss which is represented as

$$\arg\max_{(A',X')} \mathcal{L}_{atk}(A', X'; W, v_0), \tag{5.6}$$

where $A' = A \pm e$ and $X' = X \pm f$ are modified adjacency matrix and feature matrix, respectively. e is a modified edge and f is a modified feature.

To design the perturbation better, i.e., degrading the node classification model performance as much as possible, two scoring functions s_{struct} and s_{feat} are defined to evaluate the impact of the perturbation of edges and features, respectively:

$$s_{struct}(e; G, v_0) := \mathcal{L}_{atk}(A', X; W, v_0), \qquad (5.7)$$

$$s_{feat}(f; G, v_0) := \mathcal{L}_{atk}(A, X'; W, v_0). \qquad (5.8)$$

The adversarial graphs are crafted according to the scoring functions. Attackers calculate the scoring function for the entry in the candidate set of the features and edges, and then they pick the perturbation with the highest score. This process is repeated until it exceeds the budget.

Imperceptible perturbation is easy to be verified visually in computer vision, while it is hard in the graph setting, because of the discreteness and poor visibility. If only consider the budget Δ, it could be of a high risk that the adversarial examples are easily detected. Therefore, how to design an imperceptible perturbation is a new challenge in the graph adversarial attack. In particular, NETTACK only considers the perturbations preserving specific inherent properties of the input.

Graph Structure Preserving Perturbation Undoubtedly, the degree distribution is the most prominent characteristic of the graph structure and it often follows a power-law distribution for real networks. The adversarial graph will be easily found if its degree distribution changes a lot. Therefore, the adversarial graph is expected to have similar power-law behavior as the original one. Attackers utilize the likelihood ratio test to estimate the similarity of the adversarial graph degree distribution and the original one. The details of the operations are as follows:

First, estimate the scaling parameter α of the power-law distribution $p(x) \propto x^{-\alpha}$.

$$\alpha_G \approx 1 + |\mathcal{D}_G| \cdot [\sum_{d_i \in \mathcal{D}_G} \log \frac{d_i}{d_{min} - \frac{1}{2}}]^{-1}, \qquad (5.9)$$

where d_{min} denotes the minimum degree needs to be considered in the power-law test and $\mathcal{D}_G = \{d_v^G | v \in \mathcal{V}, d_v^G \geq d_{min}\}$ is a set of node degrees, where d_v^G denotes the degree of the node v in graph G. According to Eq. (5.9), one can estimate α_G and $\alpha_{G'}$ of the original graph and the adversarial graph, respectively. α_{comb} also can be estimated by using the combined samples $\mathcal{D}_{comb} = \mathcal{D}_G \cup \mathcal{D}_{G'}$.

Second, calculate the log-likelihood for the sample $\mathcal{D}_x$ according to the parameter α_x:

$$l(\mathcal{D}_x) = |\mathcal{D}_x| \cdot \log \alpha_x + |\mathcal{D}_x| \cdot \alpha_x \cdot \log d_{min} + (\alpha_x + 1) \sum_{d_i \in \mathcal{D}_x} \log d_i. \qquad (5.10)$$

Third, set up the significance test to determine whether two samples $\mathcal{D}_G$, $\mathcal{D}_{G'}$ come from the same power-law distribution (null hypothesis H_0) or not (H_1):

$$l(H_0) = l(\mathcal{D}_{comb}); l(H_1) = l(\mathcal{D}_G) + l(\mathcal{D}_{G'}). \qquad (5.11)$$

Following the likelihood ratio test, the final test statistic is

$$\Lambda(G, G') = -2 \cdot l(H_0) + 2 \cdot l(H_1), \tag{5.12}$$

which for large sample sizes follows a χ^2 distribution with one degree of freedom. The adversarial graph $G' = (A', X')$ is accepted when the p-value> 0.95 in the χ^2. Thereby, the degree distribution should fulfill

$$\Lambda(G, G') < \tau \approx 0.004. \tag{5.13}$$

Feature Statistic Preserving Perturbation It will be noticeable if two features, which never appear together in the original graph, appear together in the adversarial graph. Thus, the deterministic test is used to design the feature statistic preserving perturbation.

To select the promising feature for modifying, a co-occurrence graph $C = (\mathscr{F}, F)$ is defined for features from G, where $\mathscr{F}$ represents the set of features and $F \subseteq \mathscr{F} \times \mathscr{F}$ denotes features which appear together in the clean graph. For example, $(f_1, f_2) \in F$ means the features f_1 and f_2 co-occur in the graph. The rule of selecting the promising features is that larger the probability of reaching a feature by a random walker is, the more unnoticeable such a feature is modified. Formally, $S_u = \{j | X_{uj} \neq 0\}$ is the set of all features presenting for node u. The addition of feature $i \notin S_u$ to node u is considered unnoticeable if

$$p(i|S_u) = \frac{1}{|S_u|} \sum_{j \in S_u} \frac{1}{d_j} \cdot F_{ij} > \sigma. \tag{5.14}$$

All the perturbations must come from the graph structure preserving perturbation set or the feature statistic preserving perturbation set.

5.3.1.2 Meta Attack

Although the poisoning attack achieves better results than the evasion attack, the poisoning attack's bilevel optimization problem limits its wide application. Zügner et al. [6] explicitly tackled the problem by using meta learning and proposed the meta attack method. The key is that attackers turn the gradient-based optimization procedure of node classification models upside down and treat the input data as a hyperparameter to learn.

Meta attack [6] performed the adversarial attack on the classic GCN model. In the training phase, the parameters θ of the model is learned by minimizing the loss function $\mathscr{L}_{train}$ (e.g., cross-entropy):

$$\theta^* = \arg \min_{\theta} \mathscr{L}_{train}(f_\theta(G)), \tag{5.15}$$

where $f_\theta(\cdot)$ denotes the model function. The goal of the attackers is to minimize the attacker's loss $\mathcal{L}_{atk}$:

$$\min_{\hat{G} \in \Psi(G)} \mathcal{L}_{atk}(f_{\theta^*}(\hat{G})), \tag{5.16}$$

where $\Psi(G)$ indicates the space of the perturbations on G, and G and $\hat{G}$ represent the original graph and the adversarial one, respectively. The bilevel optimization problem can be defined as

$$\min_{\hat{G} \in \Psi(G)} \mathcal{L}_{atk}(f_{\theta^*}(\hat{G})) \quad s.t. \quad \theta^* = \arg\min_\theta \mathcal{L}_{train}(f_\theta(\hat{G})). \tag{5.17}$$

Zügner et al. introduced two attacker's loss function, and the details are as following. Attackers cannot optimize the test loss to craft adversarial graphs since the labels of the test set are not available. Given that a model will have a poor generalization performance when it has a high training error in the training phase, Zügner et al. [6] adopted the training loss $\mathcal{L}_{train}$ to solve the problem that one can not treat the test loss as the attacker's loss, that is, let $\mathcal{L}_{atk} = -\mathcal{L}_{train}$, this method is named as meta-train attack.

On the other hand, attackers learn the model by using the training set and predict the label c_u of the unlabeled node v_u. And then they can calculate the attacker's loss $\mathcal{L}_{atk}$ by using the test set and the predicted labels, that is $\mathcal{L}_{atk} = -\mathcal{L}_{self}$, which is only used to evaluate the generalization loss after training. This method names as meta-self attack.

After instantiating the attacker's loss function, Zügner et al. then used meta gradient method to handle the bilevel optimization problem, treating the graph structure matrix as a hyperparameter and calculate the gradient of the $\mathcal{L}_{atk}$ after training:

$$\nabla_G^{meta} := \nabla_G \mathcal{L}_{atk}(f_{\theta^*}(G)) \quad s.t. \quad \theta^* = opt_\theta(\mathcal{L}_{train}(f_\theta(G))), \tag{5.18}$$

where $opt(\cdot)$ is a differentiable optimization procedure. One instantiates $opt(\cdot)$ with vanilla gradient descent with learning rate α starting from some initial parameters θ_0. The parameter θ_{t+1} can be calculated as following:

$$\theta_{t+1} = \theta_t - \alpha \nabla_{\theta_t} \mathcal{L}_{train}(f_{\theta_t}(G))). \tag{5.19}$$

The meta-gradient can be expressed by unrolling the training procedure:

$$\nabla_G^{meta} = \nabla_G \mathcal{L}_{atk}(f_{\theta_T}(G)) = \nabla_f \mathcal{L}_{atk}(f_{\theta_T}(G)) \cdot [\nabla_G f_{\theta_T}(G) + \nabla_{\theta_T} f_{\theta_T}(G) \cdot \nabla_G \theta_T], \tag{5.20}$$

where $\nabla_G \theta_{t+1} = \nabla_G \theta_t - \alpha \nabla_G \nabla_{\theta_t} \mathcal{L}_{train}(f_\theta(G))$, $\mathcal{L}_{atk}(f_{\theta_T}(G)))$ denotes the attacker's loss after training for T steps.

Since the parameter θ_t is related to the initial parameter θ_0 and changes with the graph structure, therefore, the attackers can modify the graph structure according to the meta-gradient and then perform a meta update M on the data to minimize the attacker's loss function $\mathcal{L}_{atk}$:

$$G^{(k+1)} \leftarrow M(G^{(k)}). \tag{5.21}$$

The final adversarial sample $G^{(K)}$ is obtained after K rounds of attacks. Here, a straightforward way to instantiate M is meta gradient descent with some step size β, that is $M(G) = G - \beta\nabla_G\mathcal{L}_{atk}(f_{\theta_T}(G))$.

Similar to common gradient attack strategies, the meta attack method is subject to two limitations: (1) the discrete of the graph; (2) the large solution space. In order to solve the above problems, a greedy method is adopted. Attackers can use graph structure matrix A to replace the parameter G in the meta gradient formulation, since the feature matrix is fixed. What's more, a score function S is defined to estimate the modification impact of the graph $S(u, v) = \nabla^{meta}_{a_{u,v}} \cdot (-2 \cdot a_{uv} + 1)$, where a_{uv} is the entry at the position (u, v) in the adjacency matrix A. Attackers greedily pick the perturbation $e' = (u', v')$ with the highest score at a time, that is

$$e' = \underset{e=(u,v):M(A,e)\in\Psi(G)}{\arg\max} S(u, v), \tag{5.22}$$

where $M(A, e) \in \Psi(G)$ ensures that the changes performed are compliant with the attack constraints. Repeat the above operations until modified edges exceeding the budget. Finally, the adversarial graphs can be obtained.

5.3.1.3 Experiment of Results

We evaluate the performance of the above attack methods on the Cora dataset. Cora dataset consists of 2708 machine learning papers, which is one of the most popular datasets in graph data mining. Each node in the network represents a paper. All nodes are divided into seven categories: the paper based on the case, genetic algorithm, neural network, probability method, reinforcement learning, rule learning, and theory. As shown in Fig. 5.1, nodes of the same category have the same color. Every paper references at least one other paper or is referenced by other papers, which constitutes the edges. Cora dataset contains a total of 5278 edges. We perform the NETTACK and meta attack on the Cora dataset, As illustrated in Fig. 5.2, the target node is marked with a red box in the subgraphs. The prediction result of the original graph is green while it is blue after adding three edges according to the NETTACK or meta attack method. The results show that, both the two attack methods successfully fool the classifier model. The difference is that NETTACK method aims to change the classifier performance on the target node, while the meta attack method reduces the overall classification performance of the model.

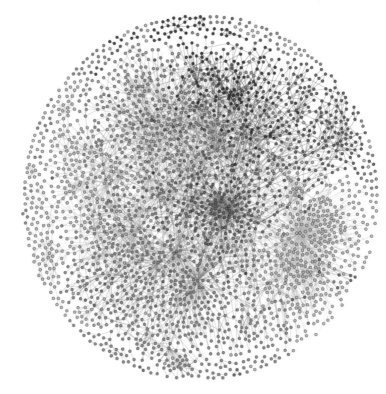

Fig. 5.1 Visualization of Cora dataset

5.3.2   Link Prediction

Link prediction is another important topic in graph data mining. For example, link prediction can be utilized to improve the success rate of biochemical experiments and reduce the cost of experiments in biology, and also recommend friends and products in social networks. However, recent studies suggest that link prediction is also vulnerable under adversarial attacks. Link prediction can also be used for friend recommendations and personalized product recommendations. Recently, the studies on link prediction adversarial attack also emerge. Waniek et al. [27] and Yu et al. [22] proposed heuristic attacks to craft adversarial graphs by rewiring links, respectively. Chen et al. [9] put forward a novel IGA method based on a graph auto-encoder framework. On dynamic network link prediction (DNLP), Chen et al. [10] mainly focused on the gradient method, and Fan et al.[19] adopted reinforcement learning, both of which can degrade the model performance. Here we mainly introduce two attack strategies, (1) The heuristic strategy proposed by Yu [22]; (2) The gradient-based strategy proposed by Chen [9].

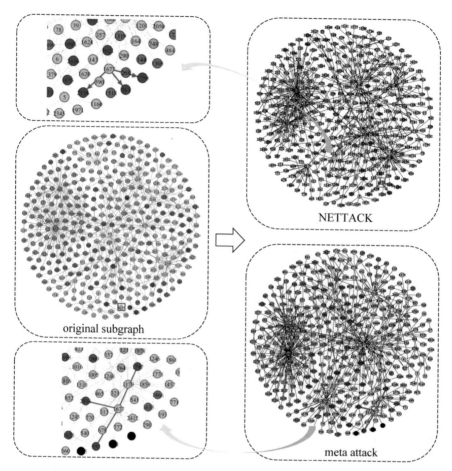

Fig. 5.2 The result of different attack methods on Cora dataset: (**a**) the NETTACK subgraph and (**b**) the meta attack subgraph. The target node has been marked with a red box in the subgraphs, and the red lines represent the added edges

5.3.2.1 Heuristic Attack

Although heuristic attack strategies may not be as effective as other optimization-based methods, they can usually reduce the computational complexity. Yu et al. [22] proposed a heuristic attack strategy based on RA index, a metric to measure the similarity between pairwise nodes, which stands out from several local similarity-based indices. It is defined as

$$RA_{ij} = \sum_{k \in \Gamma(i) \cap \Gamma(j)} \frac{1}{d_k}, \tag{5.23}$$

where $\Gamma(i)$ denotes the one-hop neighbors of node i and d_k denotes the degree of node k. In order to decrease the RA value of the target node pair, one can reduce the number of common neighbors or increase the degree of their common neighbors. Accordingly, Yu et al. [22] proposed the heuristic attack strategy and the details of the method are as following.

First, classify all the node pairs into three cases, (1) node pairs for training, (2) node pairs for test, and (3) node pairs which have no edges between them. Then, calculate each node pair's RA index and sort them in descending order according to their RA values. Finally, one can traverse the ordered node pairs and conduct different operations by deleting or adding edges for each case. The rules of each case are as follows:

- **Node pairs for training:** Delete the edge directly. In the training phase, the high RA value in the existing edge can help the model have a promising performance. If the edge in the node pairs for training is deleted, the high RA value of the target node pairs, which have no edge, will fool the model and destroy the model performance.
- **Node pairs for test:** Considering the node pair (i, j), there are two operations for it: select the common neighbor node k with the smallest degree, and then delete the edge (i, k) or (j, k); Or select the first two nodes k and l among the common neighbor sequence with increasing degree, and add the edge (k, l) into training set. The goal of the above operations is to reduce the RA value of the edge (i, j).
- **Node pairs without edges:** From the set, in which the nodes are in the neighbor set of node i but not in the neighbor set of node j, select a node k with the smallest degree and add the edge (j, k). The purpose of this procedure is to increase the RA value of the node pairs without edges.

The above operations are shown in Fig. 5.3. One can craft the adversarial examples by conducting the above operations.

5.3.2.2   Gradient-Based Attack

Intuitively, gradient-based attack methods are simple but effective. The basic idea is to fix the trained model and regard the input as a hyperparameter to optimize.

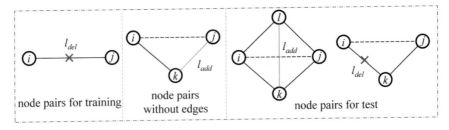

Fig. 5.3 The overview of the heuristic attack strategy

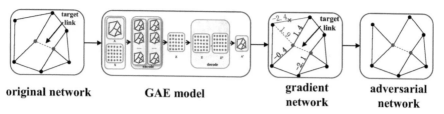

original network   GAE model   gradient network   adversarial network

Fig. 5.4 The framework of the gradient-based method

Similar to the training process, attackers could use the partial derivative of $\mathscr{L}_{atk}$ to craft the adversarial graph. Figure 5.4 shows the framework of the gradient-based method, and the details of the method are as follows:

GAE Model for Link Prediction Before the attack, the classical GAE model can be trained to have a good performance. The GAE model contains two parts, encoding model and decoding model. The encoding model adopts the classical GCN model, merges the nodes' information, and gets the embedding vector-matrix $Z \in R^{N \times F}$ of each node:

$$Z(A) = \hat{A}\sigma(\hat{A}I_N W_{(0)})W_{(1)}, \tag{5.24}$$

where A is the adjacency matrix of the graph, $\hat{A} = \tilde{D}^{-1/2}(A + I_N)\tilde{D}^{-1/2}$ is the normalized adjacency matrix, I_N is the identity matrix, and $\tilde{D}_{ii} = \sum_j (A + I_N)_{ij}$ is the diagonal degree matrix. $W_{(0)} \in R^{N \times H}$ and $W_{(1)} \in R^{H \times F}$ represent the weight matrices of the first and second layers of GCN, respectively. N represents the number of nodes, and H and F represent the feature dimensions of the first and the second layers of GCN, respectively. σ is the ReLu activation function. The decoding model calculates the similarity of each node pair:

$$\tilde{A} = s(ZZ^T), \tag{5.25}$$

where s is the sigmoid function, and $\tilde{A} \in R^{N \times N}$ is the score matrix. The values of all elements in $\tilde{A}$ are between 0 and 1. We set the threshold to 0.5, which is used to determine the link prediction results. If the score of an edge is larger than the threshold, we consider the edge exists in the graph.

Gradient Extraction The adversarial sample is the original clean graph with perturbations generated according to the extracted gradient from GAE. For the target edge E_t, we construct the target loss function as

$$\mathscr{L}_{atk} = -\omega Y_t ln(\tilde{A}_t) - (1 - Y_t)ln(1 - \tilde{A}_t), \tag{5.26}$$

where $Y_t \in \{0, 1\}$ is the ground-truth label of target edge E_t and $\widetilde{A}_t$ is the probability of the existence of E_t calculated by GAE. According to this loss function, one can calculate the partial derivative of $\mathcal{L}_{atk}$ to adjacency matrix A, which can be represented as

$$g_{ij} = \frac{\partial \mathcal{L}_{atk}}{\partial A_{ij}}. \tag{5.27}$$

The gradient matrix is usually not symmetrical. Here, for an undirected graph, the attacker will only keep the upper triangular matrix after symmetrizing

$$\widehat{g}_{ij} = \widehat{g}_{ji} = \begin{cases} \frac{g_{ij}+g_{ji}}{2} & i < j \\ 0 & \text{otherwise.} \end{cases} \tag{5.28}$$

Generate Adversarial Graph The gradient matrix's value could be positive or negative, and the positive/negative gradient means that the direction of maximizing the target loss is to increase/decrease the value in the corresponding position of the adjacency matrix. However, gradients could not be applied directly to the input data due to the discreteness of graphs. Instead, attackers will choose the greatest absolute gradient and manipulate it to a proper value. The selection of these edges depends on the gradient's magnitude, which indicates how significant the edge affects the loss function. The larger the magnitude is, the more significantly the edge affects the target loss. Due to the discreteness of graph data, no matter how large the magnitude is, attackers cannot modify the existing/nonexistent edge whose gradient is positive/negative. These edges are regarded as not attackable, and one needs to skip them for the next attackable edge.

5.3.2.3   Experiment of Results

To evaluate the performance of the attack method, we use the well-known Cora dataset. The network contains 2708 nodes and 5208 edges, which is too large to visualize accurately. Therefore, we use a subgraph, which contains the target node pair and the modified edges, to show the details of adversarial attack. As shown in Fig. 5.5, the two graphs denote the original subgraph and the adversarial subgraph, respectively. The red nodes in the subgraphs are the target node pair, and the red edges are added according to the gradient matrix. The link prediction result shows that after adding two inconspicuous edges, the edge exists in the original network will be predicted incorrectly. We randomly choose 100 edges for testing, and the results show that modifying only ten edges on average will make the link prediction model have a bad prediction performance.

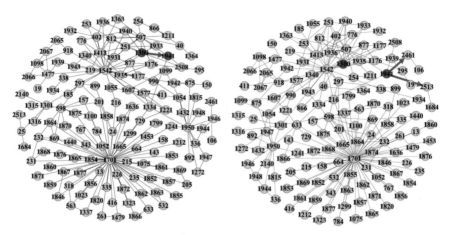

Fig. 5.5 The result of gradient-based attack method for link prediction on the Cora dataset. The left and right graphs indicate the subgraphs before and after the attack, respectively. The two red nodes make up the targeted node pair, and the red edges are added according to gradients of the adjacency matrix

5.3.3   Graph Classification

Graph classification is a task which aims to predict the class label for an entire graph. It has wide applications in different areas, for example, fraud detection [28–31], malware detection [32–35], and protein analysis [36]. There are few studies about the adversarial attacks on graph classification. For instance, Dai et al. [17] used reinforcement learning to attack GNNs by rewiring edges. Zhang et al. [37] and Xi et al. [38] proposed backdoor-based methods and achieved good performance. In this section, we will mainly introduce reinforcement learning attack in detail.

5.3.3.1   Hierarchical Reinforcement Learning Attack

The attack method based on reinforcement learning proposed by Dai et al. [17] is not only effective for graph classification, but also can destroy node classification tasks. Here, we mainly introduce the attack method against graph classification tasks. Dai et al. [17] selected edges for adding by leveraging reinforcement learning. The selection process is modeled as a Finite Horizon Markov Decision Process (MDP) as shown in Fig. 5.6 and the details of MDP are as follows:

- **Action:** For each time step t, a single action is denoted as $a_t \in V \times V$, which is viewed as the manipulating edge in the graph, where V denotes the nodes set.
- **State:** The state s_t at time t is denoted by $\hat{G}$, and $\hat{G}$ is an adversarial graph after executing the action a_t.

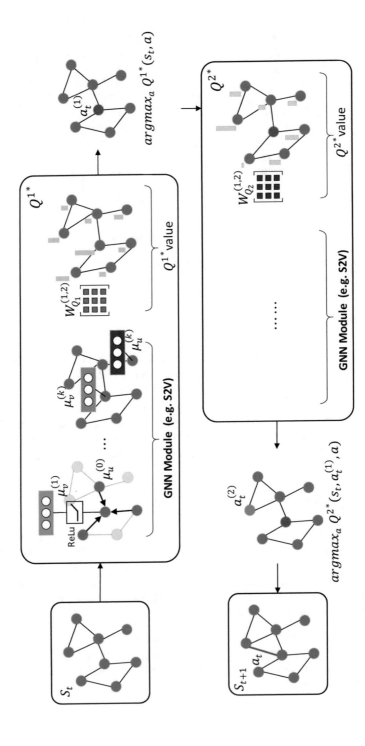

Fig. 5.6 Illustration of hierarchical reinforcement learning attack [17]

- **Reward:** Reward conduces to attackers to fool the link prediction model, so one can receive the non-zero reward at the end of MDP, and the reward is defined as follows:

$$r(\hat{G}) = \begin{cases} 1 & f(\hat{G}) \neq y \\ -1 & f(\hat{G}) = y. \end{cases} \tag{5.29}$$

Note that no reward can be received in the intermediate steps of MDP. In addition to this reward, attackers can also use classifier loss $\mathcal{L}(\hat{G}, y)$ as a reward.
- **Terminal:** The process keeps running until m edges are modified. If fewer edges are needed to modify, the procedure won't stop working, and it will modify dummy edges.

This process starts with the original graph, i.e., $s_1 = G$. The whole trajectory can be represented by $(s_1, a_1, r_1, \cdots, s_m, a_m, r_m, s_{m+1})$, and s_{m+1} is the end state, denoting an adversarial graph with m modification. Note that the last step will receive the reward r_m, and all other intermediate rewards are zero, $r_t = 0, \forall t \in \{1, 2, \ldots, m - 1\}$. Besides, Q-learning is used to learn MDP since the process is a discrete optimization problem with a finite horizon.

Q-learning is an off-policy optimization and fits the Bellman optimization equation as below:

$$Q^*(s_t, a_t) = r(s_t, a_t) + \max_{a'} Q^*(s_{t+1}, a'). \tag{5.30}$$

This implicitly suggests a greedy policy:

$$\pi(a_t|s_t; Q^*) = \arg\max_{a_t} Q^*(s_t, a_t). \tag{5.31}$$

Obviously, selecting a node pair needs $O(|V|^2)$ complexity which is too expensive. To handle this problem, the action is decomposed into $a_t = (a_t^{(1)}, a_t^{(2)})$, where $a_t^{(1)}, a_t^{(2)} \in V$. That is, one action is split into two steps, i.e., selecting two nodes respectively. The hierarchical Q-function is then modeled as below:

$$Q^{*(1)}(s_t, a_t^{(1)}) = \max_{a_t^{(2)}} Q^{*(2)}(s_t, a_t^{(1)}, a_t^{(2)}) \tag{5.32}$$

$$Q^{*(2)}(s_t, a_t^{(2)}, a_t^{(2)}) = r(s_t, a_t = (a_t^{(1)}, a_t^{(2)})) + \max_{a_t^{(1)}} Q^{*(1)}(s_t, a_{t+1}^{(1)}). \tag{5.33}$$

In this case, the computational complexity is decreased to $O(|V|)$.

The above mentioned process is for one graph. However, in graph classification, a dataset containing N graphs $\mathbf{D} = (G_i, y_i)_{i=1}^N$, which means N MDPs should be designed. In order to solve the problem caused by high computation complexity, Dai et al. [17] introduces a more practical and challenging setting: just let one Q^* be learned, and transfer it to all the MDPs.

$$\max_{\theta} \sum_{i=1}^{N} \mathbb{E}_{t,a=\arg\max_{a_t} Q^*(a_t|s_t;\theta)}[r(\hat{G}_i)], \tag{5.34}$$

where Q^* is parameterized by θ. Then present the parameterization for Q^* that generalizes over all MDPs.

The most flexible parameterization would be implementing $2N$ time-dependent Q function, while two distinct parameterizations are enough for the challenging setting. For example, at every time step, $Q_t^{*(1)} = Q^{*(1)}$, $Q_t^{*(2)} = Q^{*(2)}$. Attackers get a generalized Q function to score the node by leveraging the GNN model, and the Q function parameterized by GNN model is represented as

$$Q^{*(1)} = W_{Q_1}^{(1)} \sigma(W_{Q_1}^{(2)\top}[\mu_{a_t^{(1)}}, \mu(s_t)]), \tag{5.35}$$

where $\mu_{a_t^{(1)}}$ represents node embedding vector of node $a_t^{(1)}$ at the state s_t and $\mu(s_t) = \sum_{v \in \hat{V}} \mu_v$. $Q^{*(2)}$ can be calculated by

$$Q^{*(2)}(s_t, a_t^{(1)}, a_t^{(2)}) = W_{Q_2}^{(1)} \sigma(W_{Q_1}^{(2)\top}[\mu_{a_t^{(1)}}, \mu_{a_t^{(2)}}, \mu(s_t)]). \tag{5.36}$$

5.3.3.2 Experimental Result

To test the effect of reinforcement learning attack, we generate 15,000 synthetic graphs by utilizing Erdos-Renyi random graph model, and label them according to the structural characteristics of the graphs. We take the random attack method as the baseline and conduct experiments on graphs of different sizes. The experimental results are shown in Table 5.1. It can be seen that the reinforcement learning attack is more effective than the random attack. Besides, as the graph size becomes larger, the attack success rate decreases under the same restrictions.

Table 5.1 The results of the clean graphs and adversarial graphs crafted by random attack and reinforcement learning attack

Attack method	15–20 nodes	40–50 nodes	90–100 nodes
(Unattacked)	94.67%	94.67%	96.67%
Rand attack	78.00%	75.33%	69.33%
RL-S2V	44.00%	58.67%	62.67%

5.3.4 Community Detection

Community detection has drawn great attention nowadays. With the rapid development of community detection algorithms in recent years, a new challenging problem arises, i.e., information over-mined. People realize that some privacy information will be over-mined by those social network analysis tools. In order to protect our privacy information from being over-mined, we can use adversarial attack methods to degrade the deep model performance. There are several adversarial attack methods [15, 39, 40], and then we will introduce one of them, namely Q-Attack, in detail [15].

5.3.4.1 GA-Based Q-Attack

Genetic algorithm is a typical optimization algorithm that has been widely used in reality. It is mainly designed according to the genetic process in biology. Inspired by the principle of survival and inferiority in Darwin's biological theory, genetic algorithms obtain relatively good solution by using selection, crossover, mutation and other operations. The genetic algorithm can search for the optimal solution globally and avoid falling into local optimal solutions. Moreover, the genetic algorithm has a large expansibility, and it is easy to combine other algorithms to solve problems. With its inherent parallelism, it can conveniently perform distributed computing. Before introducing the attack method, we first introduce the coding method and fitness function, which will greatly affect the computational complexity

- **Coding:** Each individual consists of two chromosomes, i.e., edges for deleting or adding. The length of theses chromosomes must be consistent because attackers craft adversarial graphs by rewiring edges, which ensures that the graph size is not changed.
- **Fitness Function:** Modularity is an important evaluation to assess the quality of a particular partition for the graph. Here, the fitness function as is defined as:

$$f = e^{-Q}, \tag{5.37}$$

which indicates that individuals of lower modularity will have larger fitness.

The overview of the GA-based Q-attack is shown in Fig. 5.7. In the following, a specific illustration of a community detection attack based on GA is given as follows.

- **Initialization:** A initial population of fixed size is randomly generated. Note that attackers should avoid the conflict of deleting edges or adding edges, i.e., do not delete or add an edge repeatedly. Each individual represents a solution to attacks.

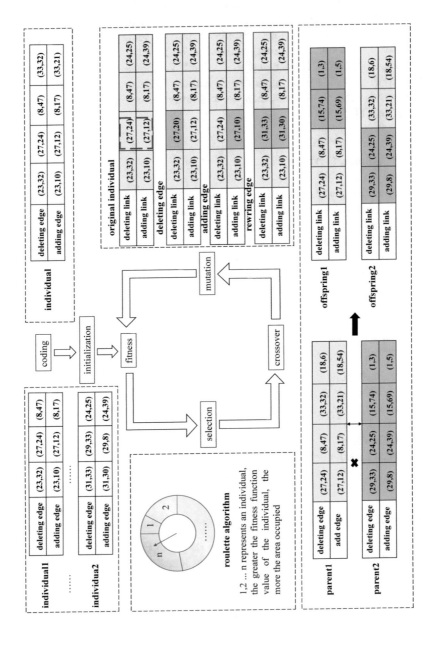

Fig. 5.7 The framework of the GA-based Q-attack

- **Selection:** Roulette wheel is adopted as a selection method. Therefore, the probability of the individual is proportional to its fitness and is represented by

$$p_i = \frac{f(i)}{\sum_{j=1}^{n} f(j)}.$$ (5.38)

- **Crossover:** Attackers generate new solutions by arranging and combining the existing gene. The one-point crossover method, as the most simple crossover method, is used in the attack, i.e., creating a breakpoint randomly and then exchanging the genes after the breakpoint to obtain two offsprings.
- **Mutation:** Mutation try to solve the problem of falling into local optimal solutions. There are three types of mutation operators: (1) deleting edge; (2) adding edge; (3) rewiring edges.
- **Elitism:** Over the course of evolution, individuals with high fitness in the parent may be destroyed. Elitism strategy has been proposed to solve this problem. For example, attackers can replace the worst 10% of the offsprings with the best 10% of the parents to maintain excellent genes.
- **Termination Criteria:** Set the evolutionary generation as a constant and the algorithm stops when the condition is fulfilled.

In summary, GA-based Q-Attack mainly involves the encoding, fitness function, and the design of genetic operators.

5.3.4.2 Experiment Result

We first briefly introduce the metrics for the community structure used in the experiment.

- **Modularity:** It is widely used to measure the quality of divisions of a graph, especially for graphs with unknown community structure. Modularity Q measures the difference between the actual fraction of within-community edges and the expected value of the same quantity with random connections and it is mathematically defined as:

$$Q = \sum (e_{ii} - a_i^2),$$ (5.39)

where e_{ii} represents the fraction of edges with two terminals both in cluster C_i, $a_i = \sum_j e_{ij}$ represents the fraction of edges that connect to the nodes in cluster C_i.
- **Normalized Mutual Information (NMI):** It is another commonly used criterion to assess the quality of clustering results in analyzing network community structure. The value of NMI indicates the similarity between the two partitions. When NMI is equal to 1, the two partitions are the same. For two partitions X

and Y, the mutual information $I(X, Y)$ is defined as the relative entropy between the joint distribution $P(XY)$ and the production distribution $P(X)P(Y)$

$$I(X, Y) = D(P(X, Y)) || P(X)P(Y) = \sum_{x,y} p(x, y) log \frac{p(x, y)}{p(x)p(y)}. \quad (5.40)$$

A noticeable problem for mutual information alone being a similarity measure is that subpartitions derived from Y by splitting some of its clusters into small ones would have the same mutual information with Y. Thus, NMI is proposed to deal with this problem and is defined by

$$I_{norm}(X, Y) = \frac{2I(X, Y))}{H(X) + H(Y)}. \quad (5.41)$$

We perform the GA-based attack method on the polbooks dataset with 105 nodes and 441 edges, which is constructed based on American politics-related books sold on Amazon. The nodes represent American politics-related books sold on Amazon online bookstores, and the edges indicate that the same consumer purchases them. Informap algorithm, which is based on network maps and coding theory, is utilized to detect the community. As shown in Fig. 5.8, the Informap algorithm is fooled by modifying 9 edges. The original graph is divided into six categories with $NMI = 0.503, Q = 0.523$ and the adversarial one is divided into eight categories with $NMI = 0.471, Q = 0.486$. The red edges in the original graph will be deleted, and those in the adversarial one will be added. As expected, the value of Q is declined because the fitness is designed to minimize Q. The value of NMI decreases, indicating that the GA-based attack method is effective. It can weaken the strength of community structure and reduce the similarity between the community detection results and the true labels.

5.4 Conclusion

Recently, graph adversarial attack has attracted considerable attention in the field of graph data mining. Driven by both the industrial and academic requests for algorithm security, there are many problems worth studying in graph adversarial attack. We summarize the main challenges as follows:

- **Unified and specified formulation:** At present, there are no clear mathematical formulas for the adversarial attack on graphs. Most of the existing researches do not have clear explanations, which makes subsequent researches more difficult.
- **Related evaluation metrics:** The evaluation system of the adversarial attack is not yet sound. Most researches use attack success rate and perturbation size to evaluate the performance of the adversarial attack. These two metrics can not fully evaluate the attack performance. Thus, more metrics are needed in the future to evaluate the performance of the attack strategy.

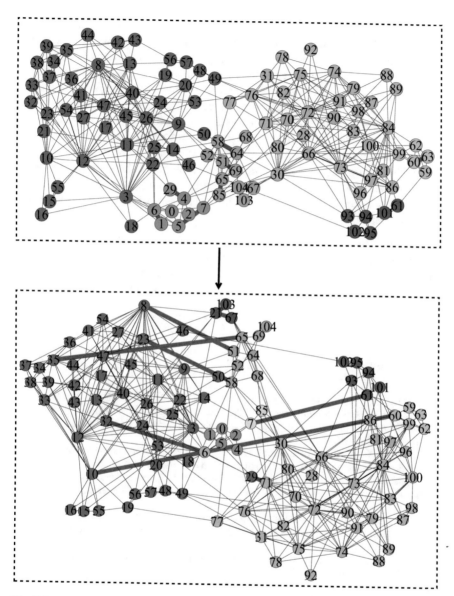

Fig. 5.8 The result of the GA-based Q-attack on polbooks dataset. The top one is the original network and the bottom one represents the adversarial one. The red edges in the original network are deleted and those on the adversarial one are added

- **Real network attack:** Existing researches rely on ideal assumptions and it is hard to apply them to real complex scenarios. For instance, we may modify the important edges to make the attack work. According to the existing attack metrics, this attack method is promising, but it could be easily detected in real applications. Besides, current attack strategies mainly focus on static networks, but a majority of real networks change over time. Therefore, we should propose more adversarial attack methods against dynamic networks in the future.

- **The definition of imperceptible:** In computer vision, it can be considered imperceptible, if one can not distinguish the difference between the original image and the adversarial one by naked eyes. Moreover, one can use some special distance functions to evaluate, such as L_p norm distance. However, in graphs, how to define *imperceptible* or *subtle perturbation* requires further investigation.
- **The interpretability of adversarial attacks:** A number of adversarial attacks have been proposed to destroy graph data mining algorithms, but we still don't understand why they work by only modifying a few edges in networks. In other words, there are few works on the interpretability of GNNs and the related adversarial attacks.

It is hoped that a sound evaluation system of adversarial attacks can be built in the future. Existing adversarial attack methods can be applied to real networks to protect personal privacy. The research on adversarial attacks can also bring some insights to better understand graph data mining methods, which could promote the further development.

References

1. Szegedy, C., Zaremba, W., Sutskever, I., Bruna, J., Erhan, D., Goodfellow, I. and Fergus, R.: Intriguing properties of neural networks. Preprint. arXiv:1312.6199 (2013)
2. Goodfellow, I.J., Shlens, J. and Szegedy, C.: Explaining and harnessing adversarial examples. Preprint. arXiv:1412.6572 (2014)
3. Carlini, N., Wagner, D.: Towards evaluating the robustness of neural networks. In: 2017 IEEE Symposium on Security and Privacy (sp), pp. 39–57. IEEE, Piscataway (2017)
4. Zügner, D., Akbarnejad, A., Günnemann, S.: Adversarial attacks on neural networks for graph data. In: Proceedings of the 24th ACM SIGKDD International Conference on Knowledge Discovery & Data Mining, pp. 2847–2856 (2018)
5. Zügner, D., Borchert, O., Akbarnejad, A., Guennemann, S.: Adversarial attacks on graph neural networks: Perturbations and their patterns. ACM Trans. Knowl. Discov. Data (TKDD) **14**(5), 1–31 (2020)
6. Zügner, D., Günnemann, S.: Adversarial attacks on graph neural networks via meta learning. Preprint. arXiv:1902.08412 (2019)
7. Chen, J., Chen, Y., Zheng, H., Shen, S., Yu, S., Zhang, D., Xuan, Q.: MGA: Momentum gradient attack on network. Preprint. arXiv:2002.11320 (2020)
8. Wu, H., Wang, C., Tyshetskiy, Y., Docherty, A., Lu, K., Zhu, L. : Adversarial examples on graph data: Deep insights into attack and defense. Preprint. arXiv:1903.01610 (2019)
9. Chen, J., Lin, X., Shi, Z., Liu, Y.: Link prediction adversarial attack via iterative gradient attack. IEEE Trans. Comput. Soc. Syst. **7**(4), 1081–1094 (2020)
10. Chen, J., Zhang, J., Chen, Z., Du, M. and Xuan, Q. Time-aware gradient attack on dynamic network link prediction. Preprint. arXiv:1911.10561 (2019)
11. Tang, H., Ma, G., Chen, Y., Guo, L., Wang, W., Zeng, B., Zhan, L.: Adversarial attack on hierarchical graph pooling neural networks. Preprint. arXiv:2005.11560 (2020)
12. Li, J., Xie, T., Chen, L., Xie, F., He, X., Zheng, Z.: Adversarial attack on large scale graph. Preprint. arXiv:2009.03488 (2020)

13. Ma, J., Ding, S., Mei, Q.: Black-box adversarial attacks on graph neural networks with limited node access. Preprint. arXiv:2006.05057 (2020)
14. Finkelshtein, B., Baskin, C., Zheltonozhskii, E., Alon, U.: Single-node attack for fooling graph neural networks. arXiv e-prints, pages arXiv–2011 (2020)
15. Chen, J., Chen, L., Chen, Y., Zhao, M., Yu, S., Xuan, Q., Yang, X.: Ga-based q-attack on community detection. IEEE Trans. Comput. Soc. Syst. **6**(3), 491–503 (2019)
16. Yu, S., Zheng, J., Chen, J., Xuan, Q., Zhang, Q.: Unsupervised euclidean distance attack on network embedding. In: 2020 IEEE Fifth International Conference on Data Science in Cyberspace (DSC), pp. 71–77. IEEE, Piscataway (2020)
17. Dai, H., Li, H., Tian, T., Huang, X., Wang, L., Zhu, J., Song, L.: Adversarial attack on graph structured data. Preprint. arXiv:1806.02371 (2018)
18. Ma, Y., Wang, S., Derr, T., Wu, L., Tang, J.: Attacking graph convolutional networks via rewiring. Preprint. arXiv:1906.03750 (2019)
19. Fan, H., Wang, B., Zhou, P., Li, A., Pang, M., Xu, Z., Fu, C., Li, H., Chen, Y.: Reinforcement learning-based black-box evasion attacks to link prediction in dynamic graphs. Preprint. arXiv:2009.00163 (2020)
20. Wang, X., Cheng, M., Eaton, J., Hsieh, C.-J., Wu, F.: Attack graph convolutional networks by adding fake nodes. Preprint. arXiv:1810.10751 (2018)
21. Chen, J., Zhang, D., Lin, X.: Adaptive adversarial attack on graph embedding via GAN. In: International Symposium on Security and Privacy in Social Networks and Big Data, pp. 72–84. Springer, Singapore (2020)
22. Yu, S., Zhao, M., Fu, C., Zheng, J., Huang, H., Shu, X., Xuan, Q., Chen, G.: Target defense against link-prediction-based attacks via evolutionary perturbations. IEEE Trans. Knowl. Data Eng. **33**(2), 754–767 (2021)
23. Zhou, K., Michalak, T.P., Rahwan, T., Waniek, M., Vorobeychik, Y.: Attacking similarity-based link prediction in social networks. Preprint. arXiv:1809.08368 (2018)
24. Bojchevski, A., Günnemann, S.: Adversarial attacks on node embeddings via graph poisoning. In: International Conference on Machine Learning, pp. 695–704. PMLR (2019)
25. Chen, J., Wu, Y., Xu, X., Chen, Y., Zheng, H., Xuan, Q.: Fast gradient attack on network embedding. Preprint. arXiv:1809.02797 (2018)
26. Xu, K., Chen, H., Liu, S., Chen, P.-Y., Weng, T.-W., Hong, M., Lin, X.: Topology attack and defense for graph neural networks: An optimization perspective. Preprint. arXiv:1906.04214 (2019)
27. Waniek, M., Zhou, K., Vorobeychik, Y., Moro, E., Michalak, T.P., Rahwan, T.: Attack tolerance of link prediction algorithms: How to hide your relations in a social network. Preprint. arXiv:1809.00152 (2018)
28. Huang, J., Xie, Y., Yu, F., Ke, Q., Abadi, M., Gillum, E., Mao, Z.M.: Socialwatch: detection of online service abuse via large-scale social graphs. In: Proceedings of the 8th ACM SIGSAC Symposium on Information, Computer and Communications Security, pp. 143–148 (2013)
29. Gong, N.Z., Frank, M., Mittal, P.: Sybilbelief: A semi-supervised learning approach for structure-based sybil detection. IEEE Trans. Inf. Forens. Secur. **9**(6), 976–987 (2014)
30. Weber, M., Domeniconi, G., Chen, J., Weidele, D.K.I., Bellei, C., Robinson, T., Leiserson, C.E.: Anti-money laundering in bitcoin: Experimenting with graph convolutional networks for financial forensics. Preprint. arXiv:1908.02591 (2019)
31. Wang, B., Jia, J., Gong, N.Z.: Graph-based security and privacy analytics via collective classification with joint weight learning and propagation. Preprint. arXiv:1812.01661 (2018)
32. Kong, D., Yan, G.: Discriminant malware distance learning on structural information for automated malware classification. In: Proceedings of the 19th ACM SIGKDD International Conference on Knowledge Discovery and Data Mining, pp. 1357–1365 (2013)
33. Nikolopoulos, S.D., Polenakis, I.: A graph-based model for malware detection and classification using system-call groups. J. Comput. Virol. Hack. Tech. **13**(1), 29–46 (2017)
34. Hassen, M., Chan, P.K.: Scalable function call graph-based malware classification. In: Proceedings of the Seventh ACM on Conference on Data and Application Security and Privacy, pp. 239–248 (2017)

35. Yan, J., Yan, G., Jin, D.: Classifying malware represented as control flow graphs using deep graph convolutional neural network. In: 2019 49th Annual IEEE/IFIP International Conference on Dependable Systems and Networks (DSN), pp. 52–63. IEEE, Piscataway (2019)
36. Hamilton, W., Ying, Z., Leskovec, J.: Inductive representation learning on large graphs. In: Advances in Neural Information Processing Systems, pp. 1024–1034 (2017)
37. Zhang, Z., Jia, J., Wang, B., Gong, N.Z.: Backdoor attacks to graph neural networks. Preprint. arXiv:2006.11165 (2020)
38. Xi, Z., Pang, R., Ji, S., Wang, T.: Graph backdoor. Preprint. arXiv:2006.11890 (2020)
39. Chen, Y., Nadji, Y., Kountouras, A., Monrose, F., Perdisci, R., Antonakakis, M., Vasiloglou, N.: Practical attacks against graph-based clustering. In: Proceedings of the 2017 ACM SIGSAC Conference on Computer and Communications Security, pp. 1125–1142 (2017)
40. Chen, J., Chen, Y., Chen, L., Zhao, M., Xuan, Q.: Multiscale evolutionary perturbation attack on community detection. Preprint. arXiv:1910.09741 (2019)

Chapter 6
Adversarial Defenses on Graphs: Towards Increasing the Robustness of Algorithms

Huiling Xu, Ran Gan, Tao Zhou, Jinhuan Wang, Jinyin Chen, and Qi Xuan

Abstract Graph Neural Networks (GNNs) have achieved tremendous development on perceptual tasks in recent years, such as node classification, graph classification, link prediction, community detection and so on. However, recent studies show that GNNs are incredibly vulnerable to adversarial attacks, so enhancing the robustness of such models remains a significant challenge. This chapter will introduce some of the latest typical defensive measures on graph-structured data to defend against malicious attacks. More specifically, we will elaborate on the existing defensive work from the following five categories: adversarial training, graph purification, certifiable robustness, attention mechanism, and adversarial detection. We will also make a comparison and summary of this existing defense methods. Different kinds of defense methods have different application scenarios and corresponding limitations.

6.1 Introduction

Recently, artificial neural network has been a hot topic in the field of artificial intelligence and made great achievements in data mining, machine translation, natural language processing, and other related fields. As a powerful form of expression, graph plays an increasingly important role in the real world and has been widely used [1]. More and more researchers have studied graph structure data through various means to excavate more value of graph, and brought many convenient and practical results to social network [2], e-commerce network [3] and recommendation system [4]. On the other hand, Graph Neural Networks (GNNs) have also attracted extensive research interest and achieved remarkable results in

H. Xu · R. Gan · T. Zhou · J. Wang · J. Chen · Q. Xuan (✉)
Institute of Cybersapace Security, Zhejiang University of Technology, Hangzhou, China

College of Information Engineering, Zhejiang University of Technology, Hangzhou, China
e-mail: xuanqi@zjut.edu.cn

© The Author(s), under exclusive license to Springer Nature Singapore Pte Ltd. 2021
Q. Xuan et al. (eds.), *Graph Data Mining*, Big Data Management,
https://doi.org/10.1007/978-981-16-2609-8_6

various graph analysis tasks such as node classification [5, 6], graph classification [7–10], etc.

Despite their remarkable performance, recent researches have clarified that GNNs are powerless against tiny but malicious attacks [11–15]. Zügner et al. [12] proposed a greedy algorithm to attack the semi-supervised node classification task. This method deliberately tries to modify the graph structure and node features such that the label of a targeted node can be changed. Dai et al. [16] proposed a reinforcement-learning based algorithm to attack both node classification and graph classification tasks by only modifying the graph structure. Further, Zügner and Günnemann [14] studied the poisoning attack based on the GNNs in node classification task, and the core of which is to use meta gradient to solve the two-layer optimization problem of poisoning attack. With the development of adversarial attack and defense on graph structured data, more and more researchers begin to pay attention to the robustness of the graph-classification model. Tang et al. [17] explored the Hierarchical Graph Pooling (HGP) neural networks' vulnerability and designed a surrogate model consisting of convolutional and pooling operators to generate adversarial samples and fool the hierarchical GNN-based graph classification models. RL-S2V [16] and ReWatt [18] are two reinforcement-learning based attack strategies, they both use the Markov Decision Process (MDP) as the attack strategies. Zhang et al. [19] proposed the graph based backdoor attack method inspired by the image field and introduced it into the graph classification. For more adversarial attack strategies, please see Chap. 5.

In order to defend the attacker's malicious attack and improve the robustness of the model, different promising methods of defense have been proposed. Bojchevski et al. [20] combined PageRank with the MDP to verify the features containing the GNNs and transmission structure to demonstrate the certified robustness. Zhu et al. [21] proposed a Robust GCN (RGCN) model which strengthens the GCN algorithms by adopting Gaussian distribution in the hidden layer and introduced variance-based recognition weights into neighboring nodes. Tang et al. [22] proposed a model of PA-GNN for defending against poisoning attacks, which relies on the punishment mechanism to directly limit the negative impact on the attack edge by assigning a low distribution coefficient. Moreover, combined with the optimization algorithm, the training of PA-GNN uses a clean graph and the corresponding adversarial graph to punish the disturbance, improving PA-GNN robustness against the poisoned graph. Some other works did not aim to harden or change the models but rather detected adversarial samples during operation. These adversarial detection models protect the GNN models by exploring the intrinsic difference between adversarial edges/nodes and the clean edges/nodes. Xu et al. [23] first proposed the detection approaches to find adversarial examples on graph data. Besides, the study of [24] introduced a graph-based random sampling approach to detect various anomaly generation models and adversarial attacks. Although these detection methods can effectively detect abnormal nodes, they are incompatible when facing graph-level adversarial attacks on graph classification tasks.

After careful research and exploration, we mainly divided all the existing methods of adversarial defense into the following five categories: adversarial training [25–28], graph purification [29, 30], certifiable robustness [20, 31, 32], attention mechanism [21, 22] and adversarial detection [23, 24, 33]. And we will introduce the above five types of adversarial defense in turn from Sects. 6.2 to 6.6, in which two examples are selected for each type. Then, we will make a comparative analysis of the current defense work and summarize its contribution in Sect. 6.7. In Sect. 6.8, we will show the experimental work and data analysis of the two defenses. And finally, we conclude the current chapter in Sect. 6.9.

6.2 Adversarial Training

Adversarial training (AT) has been widely used in the field of image attack and defense. The core idea of adversarial training is to mix the adversarial samples into training samples during the normal training to improve the robustness of the trained model. In recent years, several defense methods based on adversarial training have emerged. Inspired by this, some studies have introduced adversarial training into the network field and achieved good performances. In the following, we will introduce two representative adversarial training methods: Graph Adverse Training [34] and SAT [28].

6.2.1 Graph Adversarial Training

Although graph neural network has good performance, Feng et al. [34] considered that graph neural networks are vulnerable to small but intentional perturbations on the input features. Figure 6.1 gives a simple but intuitive example to illustrate how the perturbation affects the classification results of nodes and their neighbors.

Adversarial Training (AT) is a dynamic regularization method that proactively simulates the perturbations during the training phase [25]. At present, adversarial training has been proved to be able to stabilize neural networks and enhance their robustness to standard classification tasks [26, 27]. However, because the traditional adversarial training does not take into account the influence of the connected examples, it is not enough to use traditional adversarial training directly in the graph neural network training process.

Therefore, Feng et al. proposed a new adversarial training method called Graph Adversarial Training (GraphAT), which learns to resist perturbations by taking the graph structure into account during adversarial training. In essence, GraphAT can be considered as a dynamic regularization scheme based on the graph structure. They applied the GraphAT to Graph Convolutional Network (GCN) and proved that the GCN model with adversarial training has higher classification accuracy and robustness than the standard trained GCN. In the following, we first introduce

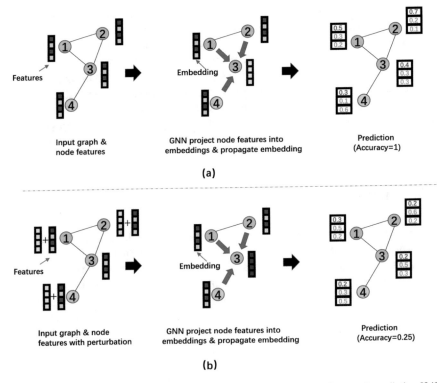

Fig. 6.1 How does perturbations to node features affect the graph neural network prediction [34]. (**a**) Standard training. (**b**) Adversarial training applying perturbations

the definition of GraphAT and then show the GraphVAT combined with virtual adversarial regularization which is considered as an extension of GraphAT.

Graph Adversarial Training (GraphAT) Inspired by the standard AT, Feng et al. developed graph adversarial training, which trains the graph neural network model by generating adversarial examples and optimizing regularization terms of adversarial examples, to prevent the adverse effects of perturbations. What makes it unique is that GraphAT aims to prevent perturbations propagating through node connections. Figure 6.2 describes the GraphAT's training process. The formula of GraphAT is:

$$
\begin{aligned}
\min : \Gamma_{GraphAT} &= \Gamma + \beta \sum_{i=1}^{N} \sum_{j \in ne_i} d\left(f\left(x_i + r_i^{origin}, G \mid \theta \right), f\left(x_j, G \mid \theta \right) \right), \\
\max : r_i^{origin} &= \arg\max_{r_i \|r_i\| \le \xi} \sum_{j \in ne_i} d\left(f\left(x_i + r_i, G \mid \widehat{\theta} \right), f\left(x_j, G \mid \widehat{\theta} \right) \right)
\end{aligned}
$$

(6.1)

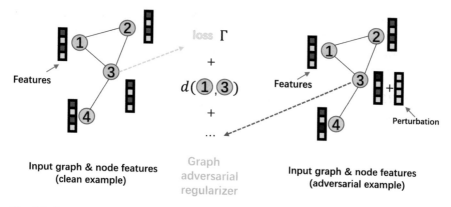

Fig. 6.2 Schematic of the GraphAT training process, with the standard objective function term on the left and the new regularization term on the right [34]

where $\Gamma_{GraphAT}$ is the training objective function with two terms: (1) standard objective function of the model based on the original graph and (2) graph adversarial regularization. The second term encourages the graph adversarial examples to be classified similarly as connected examples, where θ denotes the model parameters to be learned, and d is a nonnegative function that measures the divergence (e.g., Kullback-Leibler divergence [35]) between two predictions. x_i denotes the features of node i, r_i^{origin} denotes the adversarial perturbation of the original graph, which is established by perturbing the features of the node i.

The graph adversarial perturbation is calculated by maximizing the graph adversarial regularization under the current value of model parameters. ξ is a hyperparameter controlling the magnitude of perturbations. Usually, a smaller value is used to make the feature distribution of the adversarial sample close to that of the normal sample.

Calculating the perturbation r_i^{origin} is not an easy task. Inspired by the linear approximation method proposed in [25] for standard adversarial training, Feng et al. also designed a linear approximation method to calculate the graph adversarial perturbations in GraphAT, of which the formula is:

$$r_i^{origin} \approx \xi \frac{\mathbf{g}}{\|\mathbf{g}\|}, \text{ where } \mathbf{g} = \nabla_{x_i} \sum_{j \in ne_i} D\left(f\left(x_i, G \mid \widehat{\theta}\right), f\left(x_j, G \mid \widehat{\theta}\right)\right), \quad (6.2)$$

where $\mathbf{g}$ is the gradient of the input x_i. Since the gradient of the GNN model can be effectively calculated through a back-propagation, such approximate calculation provides a more efficient method of perturbation calculation. $\widehat{\theta}$ is a constant set denoting the current model parameters.

Virtual Graph Adversarial Training (GraphVAT) Feng et al. further devised an extended version of GraphAT (GraphVAT), which is inspired by the idea of virtual adversarial training[36]. The formula of GraphVAT is:

$$\min : \Gamma_{GraphVAT} = \Gamma + \alpha \sum_{i=1}^{N} d\left(f\left(x_i + r_i^v, G \mid \theta\right), \tilde{y}_i\right)$$

$$+ \beta \sum_{i=1}^{N} \sum_{j \in ne_i} d\left(f\left(x_i + r_i^{origin}, G \mid \theta\right), f\left(x_j, G \mid \theta\right)\right),$$

$$\max : r_i^v = \arg \max_{r_i' \| r_i' \| \leq \xi'} d\left(f\left(x_i + r_i', G \mid \hat{\theta}\right), \tilde{y}_i\right), \tag{6.3}$$

where r_i' denotes the virtual adversarial perturbation, the direction that leads to the largest change on the model prediction of x_i. For labeled nodes and unlabeled nodes, $\tilde{y}_i$ denotes truth label and model prediction, as shown below:

$$\tilde{y}_i = \begin{cases} \hat{y}_i, & i \leq M \text{ (labeled node)}, \\ f\left(x_i, G \mid \hat{\theta}\right), & M < i \leq N \text{ (unlabeled node)}. \end{cases} \tag{6.4}$$

During the GraphVAT train process, in each iteration, two types of perturbations and the associated adversarial examples are generated: (1) the smoothness of prediction around the individual clean example and (2) the smoothness of connected examples.

For the consideration of efficiency, they calculated r_i^v via the power iteration approximation as blow.

$$r_i^v \approx \xi' \frac{g}{\|g\|}, \text{ where } g = \nabla_{r_i} d\left(f\left(x_i + r_i, G \mid \hat{\theta}, \tilde{y}_i\right)\right)\big|_{r_i = \xi d}, \tag{6.5}$$

where d is a random vector. Detailed derivation of the method is referred to [36].

The advantage of GraphAT is that it does not affect the convergence speed of the GCN training process. At the same time, the model after adversarial training has smoother predictions of graph structure, so it has stronger generalization ability and robustness.

6.2.2 SAT

Most of the current algorithms based on adversarial training only focus on global defense through global adversarial training. At the same time, it is still a challenge for the existing adversarial training methods to defend against the attack of target nodes. Therefore, Chen et al. [28] proposed smooth adversarial training to improve the robustness of GNN.

We adopted their adversarial training method to enhance the defensive capability of the GCN model. In particular, we proposed two adversarial training strategies: global adversarial training (Global-AT) to protect all the nodes and target-label adversarial training (Target-AT) to protect the nodes of the target label, which are introduced in the following.

Generally speaking, our work mainly includes two kinds of adversarial training methods: Global-AT and Target-AT. Besides, two smoothing strategies are proposed: Smoothing distillation (SD) and Smoothing cross-entropy (SCE). These methods are described in detail below.

Global Adversarial Training (Global-AT) The process of Global-AT is to generate the adversarial network iteratively. The adversarial links are selected by adversarial attack methods on the training node set $S_{train} = [v_1, \cdots, v_m]$. In particular, for all nodes in S_{train}, generate the adjacency matrix $\hat{A}^t$ iteratively according to the following steps:

- **Selecting the adversarial links.** First, the adversarial attack method is used to select adversarial links to maximize the negative cross-entropy loss of the training decoder in S_{train}. These adversarial links will be saved as a matrix Λ, and their size will be the same as the adjacency matrix A. Its element $\Lambda_{ij} \in \{-1, 0, 1\}$ indicates the modification of links. If $\Lambda_{ij} = 1$, add an adversarial link between nodes v_i and v_j. If $\Lambda_{ij} = -1$, delete the relationship between v_i and v_j. If $\Lambda_{ij} = 0$, the relationship between v_i and v_j will not be modified.
- **Updating adversarial network.** After generating the adversarial links matrix Λ, update the $(t-1)$th adversarial network. The specific update process is defined as:

$$\hat{A}_{ij}^t = \hat{A}_{ij}^{t-1} + \Lambda_{ij}, \tag{6.6}$$

where $\hat{A}_{ij}^t$, $\hat{A}_{ij}^{t-1}$ and Λ_{ij} denote the elements of $\hat{A}^t$, $\hat{A}^{t-1}$ and Λ, respectively.

Target-label Adversarial Training (Target-AT) Different from Global-AT, the goal of Target-AT is only to protect those nodes with specific labels. That is to say, given a target label τ_p, the attack method only generates adversarial links for the set of these labels S_{τ_p}. The remaining steps of Target-AT are no different from Global-AT.

Smoothing Distillation (SD) As suggested by Hinton et al. [37], the distilled model is trained on a transferred dataset. The labels of the set come from soft label prediction under a high-temperature model. More specifically, the proposed distillation model consists of two modules: The first module is used to label those unlabeled nodes through the normally trained model, and the obtained labels are called soft labels; the second is the distillation module, which uses the soft labels of the data instead of the ground truth to train the classifier. The proposed distillation model can effectively improve the robustness of the classifier.

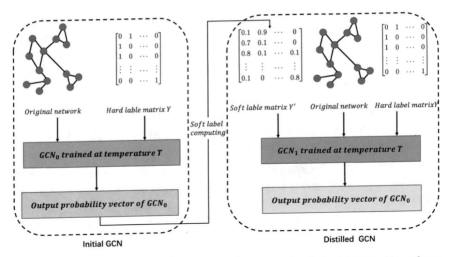

Fig. 6.3 The illustration of smoothing distillation. At first, training the initial GCN with a softmax output layer at temperature T. Then, encode training nodes by soft label. Finally, train the distilled GCN with its training objective function, the objective function is composed of soft loss function and original loss function

Inspired by the distillation model, Chen et al. proposed smoothing distillation (SD) to improve the robustness of GCN against adversarial perturbations. Compared with the original GCN classification, our distillation GCN model has higher classification accuracy and is beneficial for transplantation. At the same time, the information extracted by distillation will help filter the perturbations deliberately added to the network, thereby improving the robustness of the model. The SD framework is shown in Fig. 6.3. Different from the original distillation method, SD maintains the same model structure of GCN model and distillation model. The specific training procedures are as follows:

- **Training the initial GCN model.** Given the training set $S_{train} = [v_1, \cdots, v_m]$ and its hard label matrix Y, at first training the initial GCN model with a softmax output layer at temperature T to obtain Y' as the output confidence of the model.
- **Encoding nodes by soft label.** Using the soft label Y' to encode the confidence probability on the labels of all training nodes.
- **Training the distilled GCN model.** A distillation model is jointly trained based on the soft label matrix Y' and the real label matrix Y, and a new soft loss is added to the objective function. The objective function L_{all} is composed of the soft loss function L_s and original loss function L, defined as:

$$L_{all} = \frac{T^2 L_s}{T^2 + 1} + \frac{L}{T^2 + 1}, \tag{6.7}$$

with

$$L_s = - \sum_{l=1}^{|S_{train}|} \sum_{k=1}^{K} Y'_{lk} \ln(Y''_{lk}), \qquad (6.8)$$

where Y'' denotes the output of the distilled GCN model. Note that, in the soft loss L_s, we also use a softmax function with temperature T.

Smoothing Cross-Entropy Inspired by the model regularization method [38], we further proposed a Smoothing Cross-entropy Loss (SCEL) function. SCEL encourages the GCN model to return high confidence values for the true labels while giving each node a smooth distribution of confidence values on the wrong labels. The SCEL function is defined as:

$$L_{smooth} = - \sum_{l=1}^{|S_{train}|} \sum_{k=1}^{K} \hat{Y}_{lk} \ln(Y'_{lk}), \qquad (6.9)$$

where $\hat{Y}$ is a smoothing matrix with $\hat{Y}_{lk} = 1$ if node v_l belongs to category τ_k and $\hat{Y}_{lk} = \frac{1}{K}$ otherwise, $C = [\tau_1, \cdots, \tau_K]$ is the category set for the nodes in the network, K denotes the number of categories, and Y' is the output of the model.

Generally speaking, the advantage of the above defense methods is that gradient hiding can be carried out under global attack or target attack. At the same time, the defense mechanism is very flexible, the adversarial training methods and smoothing strategies can be combined according to the specific situation, but it will sacrifice some embedding performance of the original network to a certain extent.

6.3 Graph Purification

Unlike other defense methods such as adversarial training, graph purification usually aims at defending against poison attack. That is, it hopes that the attacked graph can be purified to restore the clean graph as much as possible for model retraining. Next, we will elaborate on graph purification defense through the following two specific methods: GCN-Jaccard [29] and GCN-SVD [30].

6.3.1 GCN-Jaccard

Wu et al. [29] thought that since the GCN model strongly relies on graph structure and local aggregation, GCN is vulnerable to attack. Therefore, the model trained on the attacked graph will be affected by the attack surface of the model made by the adversarial graph.

Their defense was inspired by the observation of the characteristics of current attack methods. First, modifying edges is a more effective attack method than modifying features. Furthermore, the attacker is more inclined to add edges rather than delete; Second, the more neighbors a node has, the harder it is usually to be attacked. Last, most attackers tend to connect the target node to another node with different features and labels.

Based on the above observations, Wu et al. believed that observing the feature similarity between nodes can assess the possibility of being attacked. They mainly introduced and used the Jaccard similarity score to evaluate the similarity of node features between nodes. Given two nodes u and v with n binary features, the Jaccard similarity score measures the overlap that u and v share with their features. Each feature of u and v can either be 0 or 1. The total number of each combination of features for both u and v are specified as follows.

M_{11} is the number of features where both nodes u and v have a value of 1. M_{01} is the number of features where the value of the feature is 0 in node u but 1 in node v. Similarly, M_{10} is the number of features where the value of the feature is 1 in node u but 0 in node v, and M_{00} represents the total number of features which are 0 for both nodes. The Jaccard similarity score is given as:

$$J_{u,v} = \frac{M_{11}}{M_{01} + M_{10} + M_{11}}. \tag{6.10}$$

To verify their point of view, they trained a two-layer GCN on the cora-ml dataset and to observe whether the nodes can be classified correctly with high probability. For these nodes, by observing the histogram of the Jaccard similarity score of the connected nodes before and after the FGSM attack, it was found that the number of neighbors with a lower similarity score of the target node significantly increases by the attack. What's more interesting is that other attack methods also show such characteristics, such as Nettack [12].

Based on the above findings, considering the efficiency of defense, Wu et al. proposed a simple and effective defense method. The core idea of this defense model is that a clean node is usually not connected to a node that is not similar to it.

The defense process is as follows: for the adjacency matrix A, calculating the edge similarity score between the connecting nodes in the graph, and then remove the edges with a similarity score less than a certain value. The entire defense process is shown in Fig. 6.4. The Jaccard similarity score is used to measure the similarity scores. Note that different similarity evaluation methods could be adopted for different data to get a better effect of defense.

It can achieve a good defense effect by removing only the edges with a Jaccard similarity score of 0. Besides, the time overhead of the defense mechanism is almost negligible. They used this defense mechanism in the GCN model, and the training time on the cora-ml and citeseer datasets only increased by 7.52 s and 3.79 s, respectively. Another advantage is that enabling this defense mechanism does not damage the classification accuracy of normal samples.

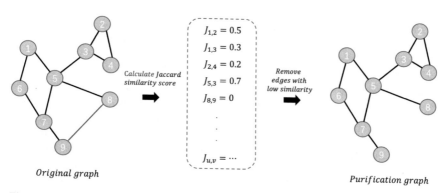

Fig. 6.4 Use the Jaccard similarity score for the defense process of the GCN model, which we call GCN-jaccard

6.3.2   GCN-SVD

In recent years, some studies show that many graph attack methods have the same characteristics. Entezari et al. [30] found that some of the attack methods, such as Nettack [12], can be considered as a high-rank attack. They found that Nettack demonstrates a very specific behavior in the spectrum of the graph: only high-rank (low-valued) singular components of the graph are affected. Therefore, they showed that a low-rank approximation of the graph, which uses only the top singular components for its reconstruction, can greatly reduce the effects of Nettack and boost the performance of GCN when facing adversarial attacks. In their study, they explored poisoning attacks on graph data and proposed a mechanism to defend against the attacks called GCN-SVD. This is another graph purification defense method.

Singular Value Decomposition (SVD) is one of the most popular matrix decomposition techniques. SVD can decompose a matrix into the sum of rank-1 matrices and is currently widely used. Let $A \in R^{I \times J}$ be a real-valued matrix, the SVD of A is computed as follows:

$$A = U \Sigma V^T, \tag{6.11}$$

where $U \in R^{I \times I}$ is the left singular matrix and the $V \in R^{J \times J}$ is the right singular matrix. $\Sigma \in R^{I \times J}$ is a non-negative diagonal matrix such that $\Sigma_{i,i} = \sigma_i$, where σ_i is the i th singular value and $\sigma_1 \geq \sigma_2 \geq \cdots \geq \sigma_{\min}(I, J)$.

The rank-r approximation of A can be computed as follows:

$$A_r = U_r \Sigma_r V_r^T = \sum_{i=1}^{r} u_i \sigma_i v_i^T, \tag{6.12}$$

where A_r is the rank-r approximation of A derived from SVD of A. U_r and V_r are the matrices containing the top r singular vectors and Σ_r is the diagonal matrix containing only the r singular values.

The perturbation imposed by Nettack can be regarded as a high-rank perturbation. By observing the singular value decomposition of the adjacency matrix of the clean graph and the adversarial graph, they found that singular values are very close at lower ranks but vary at higher ranks. This means that when we decompose the adjacency matrix, the low-rank approximation of the adversarial sample is not much different from the clean sample. To discard the high-rank perturbations, they computed the low-rank approximation of the adjacency and feature matrices derived from their SVD decomposition according to Eq. (6.12), and then retrained GCN with the low-rank approximation matrices. With a proper choice of r, the rank-r approximation of the attacked graph can boost the performance of GCN and achieve the performance close to that on the clean graph. Figure 6.5 describes how GCN-SVD uses singular value decomposition to purify the adversarial graph.

They gave the derivation that SVD uses a low-rank approximation to purify those nodes whose degree is less than a certain value. In general, the rank-r approximation can detect attacks on target nodes with a degree less than $\sigma_r^2 - 2$. Entezari et al. used different values of r to evaluate the defensive effect. Usually, $r = 10$ has better defensive performance. More proof details are available in [30].

The GCN-SVD defense model can achieve effective defense only by singular value decomposition of adjacency matrix, and does not need to adjust too many parameters. However, it will reduce the classification accuracy of normal samples to a certain extent.

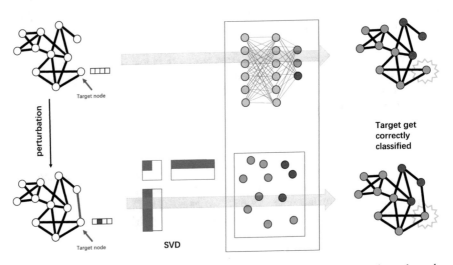

Fig. 6.5 Low-rank approximation of graph structure and feature matrices to vaccinate the node classification method and discard perturbations [30]

6.4 Robustness Certification

Robustness Certification is an effective way to defend against adversarial attacks, based on which it can be proved that certain nodes or edges will not be attacked successfully under certain perturbation, and then we can use the robustness Certification for training to improve the robustness of the model. It can prevent small changes in data from causing completely different predictions in GNN, and certify whether a given GNN is robust.

In this section, we focus on certifiable robustness. Several works proposed in [20, 31, 32] have considered the safety of GNN and tried to certify robustness. Zügner et al.[32] considered the perturbations on node attributes. Bojchevski et al.[20] dealt with the case when the attacker only change the graph structure. It derives the robustness certifications as a linear function of personalized PageRank, which makes the optimization tractable. Jia et al. [31] studied certifiable robustness on GNN's other applications such as community detection. In the rest of this section, we will focus on the above mentioned three strategies.

6.4.1 Certifying Robustness for Graph Structure Perturbations

Bojchevski et al.[20] proposed a new method for provable robustness regarding perturbations of the graph structure. Their approach is applicable to graph neural network and label/feature propagation model whose predictions are a linear function of personalized PageRank. They are the first to propose a robust certification problem in the field of node classification. In addition, they also proposed a robust training method that can simultaneously improve the robustness and accuracy of the model.

Robustness Certifications Bojchevski et al.[20] designed two strategies, one for the local budget and the other for both the local budget and the global budget. The local budget limits the number of additions and deletions to a single node, and the global budget limits the number of additions and deletions to the entire graph.

They designed an algorithm for the local budget, whose main idea is starting from a random policy: in each iteration they first compute the mean reward before teleportation for the current policy, and then greedily select the top local budget edges that improve the policy. This algorithm is guaranteed to converge to the optimal policy, and thus to the optimal configuration of fragile edges.

In view of the global budget and local budget at the same time, they developed a three-step method: (1) Based on the auxiliary graph, an unconstrained Markov decision process(MDP) alternative is proposed, which can reduce the operation set from exponential to binary level only by adding auxiliary nodes; (2) Augment the corresponding linear program with quadratic constraints to enforce the global budget; (3) Apply the Reformulation Linearization Technique (RLT) relaxation to the resulting Quadratically Constrained Linear Program (QCLP).

Robust Training Bojchevski et al. optimize the robust cross-entropy loss proposed by Wong and Kolter [39]:

$$L_{RCE} = L_{CE}\left(y_v^*, -m_{y_v}^*(v)\right),\tag{6.13}$$

where L_{CE} is the standard cross-entropy loss operating on the logits, and y_v^* is the predicted label for v, and $m_{y_v}^*(v)$ is a vector such that at index c we have $m_{y_v,c}^*(v)$, $m_{y_v,c}^*(v)$ represents the worst-case margin between class y_v^* and class c. They studied the alternative robust loss of the hinge to avoid L_{RCE}, which encourages high certainty under the worst-case perturbations. The purpose of the attacker is to minimizes the worst-case margin $m_{y_v,*}^*(v)$ (or its lower bound), so the robust training adds a hinge loss penalty term to the standard cross-entropy loss, and tries to maximize it during training:

$$L_{CEM} = \sum_{v \in V_L}\left[L_{CE}\left(y_v^*, H_{v,:}^{\text{diff}}\right) + \sum_{c \in C, c \neq y_v^*} \max\left(0, M - m_{y_v,c}^*(v)\right)\right],\tag{6.14}$$

where V_L is a subset labeled nodes, if $m_{y_v,c}^*(v) < M$ and zero, $H_{v,:}^{\text{diff}}$ is a weighted combination of the logits of all nodes, the second term for a single node v is positive, otherwise the node v is certifiably robust with a margin of at least M.

The work of Bojchevski et al.[20] has shown good results, but they only consider graph structure disturbances. In fact, many real-world disturbances are of multiple types. In the future, we could consider the robustness of both node features and graph structure disturbances on the basis of their work.

6.4.2 Certifying Robustness for Node Attributes Perturbations

Zügner et al. [32] considered the perturbation of node attributes. They asked the question: Which nodes will not be changed in the range of allowable disturbance. To answer this question, they first conducted research on graph convolutional neural networks and node attribute disturbances.

Certifying Robustness for Graph Convolutional Networks Let $G = (A, X)$ be an attributed graph, where $A \in \{0, 1\}^{N \times N}$ is the adjacency matrix and $X \in \{0, 1\}^{N \times D}$ represents the nodes' features. We assume the node-ids to be $V = \{1, \ldots, N\}$. Given a subset $V_L \subseteq V$ of labeled nodes, with class labels from $C = \{1, 2, \ldots, K\}$, the goal of node classification is to learn a function $f : V \to C$ which maps each node $v \in V$ to one class in C. Due to the particularity of the graph, the attributes of a node depend on the attributes of its surrounding nodes and edges, so the perturbations of this node are also restricted. Therefore, they sliced the adjacency matrix $A \in \{0, 1\}^{N \times N}$ and node features $X \in \{0, 1\}^{N \times D}$ to calculate the output of the target node t. The latent representations $H^{(l)}$ at layer L are of the form:

$$\hat{H}^{(l)} = \dot{A}^{(l-1)} H^{(l-1)} W^{(l-1)} + b^{(l-1)} \quad \text{for } l = 2, \ldots, L, \tag{6.15}$$

$$H^{(l)}_{nj} = \max\left\{\hat{H}^{(l)}_{nj}, 0\right\} \quad \text{for } l = 2, \ldots, L-1, \tag{6.16}$$

where $H^{(1)} = \dot{X}$. The output of this sliced GNN is denoted as $f^t_\theta(\dot{X}, \dot{A}) = \hat{H}^{(L)} \in \mathbb{R}^K$. Here θ is the set of all parameters, i.e. $\theta = \left\{W^{(\cdot)}, b^{(\cdot)}\right\}$.

According to the above operation, the maximum loss constraint can be defined: Given a graph G, a target node t, and an GNN with parameters θ. Let y and y^* denote the class of node t(e.g. given by the ground truth or predicted). In the worst case, the margin between classes y^* and y achievable under some set $X_{q,Q}(\dot{X})$ of admissible perturbations to the node attributes is given by

$$m^t\left(y^*, y\right) := \text{minimize}_{\tilde{X}} \; f^t_\theta(\tilde{X}, \dot{A})_{y^*} - f^t_\theta(\tilde{X}, \dot{A})_y$$
$$\text{subject to } \tilde{X} \in X_{q,Q}(\dot{X}), \tag{6.17}$$

where q represents the local perturbation budget, Q represents the global perturbation budget, if $m^t\left(y^*, y\right) > 0$ for all $y \neq y^*$, the GNN is certifiably robust.

Perturbations Definition According to the actual research situation in the field of graphs, they defined the permissible disturbance set by limiting the number of changes to the original attribute. The attribute of the node in the vicinity of the L-1 jump can be modified to change the prediction of the target node, and limit the number of local perturbations:

$$X_{q,Q}(\dot{X}) = \left\{\tilde{X} \mid \tilde{X}_{nj} \in \{0, 1\} \wedge \|\tilde{X} - \dot{X}\|_0 \leq Q \right.$$
$$\left. \wedge \left\|\tilde{X}_{n:} - \dot{X}_{n:}\right\|_0 \leq q \forall n \in N_{L-1}\right\}. \tag{6.18}$$

Robust Training of GNNs Zügner et al. also considered using the credential suggested in the previous section to enhance the robustness of the GNN model. In order to enhance the robustness of the model, they considered the training target normally used to train GNN for node classification and used the following methods to optimize the model:

$$\underset{\theta, \{\Omega^{t,k}\}_{t \in V_L, 1 \leq k \leq K}}{\text{minimize}} \sum_{t \in V_L} L\left(p^t_\theta\left(y^*_t, \Omega^{t,\cdot}\right), y^*_t\right) \tag{6.19}$$

where L is the cross-entropy function (operating on the logits) and V_L is the set of labeled nodes in the graph, Ω is a variable have only simple element-wise constraints (e.g. clipping between $[0, 1]$), p indicates those elements which are perturbed. To facilitate true robustness and not false certainty in models' predictions, they therefore proposed an alternative robust loss that refer to as robust hinge loss:

$$\hat{L}_M\left(\boldsymbol{p}, y^*\right) = \sum_{k \neq y*} \max\left\{0, \boldsymbol{p}_k + M\right\}. \tag{6.20}$$

This loss is positive if $-\boldsymbol{p}^t_{\theta k} = g^t_{q,Q}\left(\dot{X}, c^k, \Omega^k\right) < M$; and zero otherwise. The node t is certifiably robust if the loss is zero, even ensuring a margin of at least M to the decision boundary in this situation. Importantly, it is not rewarded to consider even greater margins (for the worst-case). They combined the robust hinge loss with standard cross-entropy to obtain the following robust optimization problem:

$$\min_{\theta, \Omega} \sum_{t \in V_L} \hat{L}_M\left(\boldsymbol{p}^t_\theta\left(y^*_t, \Omega^{t,\cdot}\right), y^*_t\right) + L\left(f^t_\theta(\dot{X}, \dot{A}), y^*_t\right), \tag{6.21}$$

$$\min_{\theta, \Omega} \sum_{t \in V_L} \hat{L}_{M_1}\left(\boldsymbol{p}^t_\theta\left(y^*_t, \Omega^{t,\cdot}\right), y^*_t\right) + L\left(f^t_\theta(\dot{X}, \dot{A}), y^*_t\right) + \sum_{t \in V \setminus V_L} \hat{L}_{M_2}\left(\boldsymbol{p}^t_\theta\left(\tilde{y}_t, \Omega^{t,\cdot}\right), \tilde{y}_t\right). \tag{6.22}$$

The above Eq. (6.21) is to train the GNN on the marked nodes until it converges, and the following equation Eq. (6.22) is to train all nodes until it converges. The robust training method they proposed is more effective than the original training method. According to the experimental results given in the paper, using robust training can increase original robustness by four times under the best conditions.

The work of Zügner et al.[32] can train a very robust model, but it can not improve the accuracy of the model. If they can consider the disturbance at the structural level at the same time, referring to the work of Bojchevski et al.[20], it may be possible to further improve the effect of robust training.

6.4.3 Certifiable Robustness in Community Detection

Community detection is a very important algorithm in the field of graphs, but it is vulnerable to adversarial structural perturbation. For example, by adding or deleting some key edges in the graph, attackers can change the results of community detection. Jia et al.[31] developed the first certifiable robustness algorithm in community detection. Given an arbitrary community detection method, they used random graph structure perturbation to construct a new smooth community detection method. Their method can prove that the predicted community of a given set of nodes will not change when the number of edges added/removed by an attacker is bounded, and they also evaluated their method on multiple real-world graphs with ground truth communities.

Attacks to Community Detection Jia et al.[31] referred to the main attack methods of current community detection algorithms[40–44] and used two typical attack methods (Splitting attack and Merging attack) to verify their certified robustness of community detection.

Computing Certified Perturbation Size Jia et al.[31] theoretically derived the certified perturbation size of the smoothed function g. Their results can be summarized into the following two theorems:

Theorem 1 *Given a graph structure binary vector x, a community detection algorithm A, and a set of victim nodes V_{victim}. Assume there exists $\underline{p} \in [0, 1]$ such that:*

$$\Pr(f(\mathbf{x} \oplus \epsilon) = y) \geq \underline{p} > 0.5, \tag{6.23}$$

where $\underline{p}$ is a lower bound of the probability $\Pr(f(\mathbf{x} \oplus \epsilon) = y)$ that f outputs y under the random noise ε, and f is a function which models the splitting and merging attacks. The function f outputs 1 if the nodes in V_{victim} are grouped into the same community detected by community detection A and outputs 0 otherwise. Pr is the noise distribution in the discrete space $\{0, 1\}^n$.

Theorem 2 *For any perturbation δ with $\|\delta\|_0 > M$(certified perturbation size), there exists a community detection algorithm A^*(and thus a function f^*) consistent with Eq. 6.23 such that $g(\mathbf{x} \oplus \delta) \neq y$ or there exists ties. For detailed derivation, please refer to the original paper.*

Based on these two theorems, they designed an algorithm to calculate the magnitude of the disturbance. Given a graph-structure binary vector x, a community detection algorithm A, and a set of victim nodes V_{victim}, the goal of the algorithm is to compute the certified perturbation size in practice. Algorithm 1 shows their complete certification algorithm. The function $SampleUnderNoise$ randomly samples N noise from the noise distribution, adds each noise to the graph structure, and computes the frequency that the function f outputs 0 and 1, respectively. Then, their algorithm estimates y^* and $\underline{p}$, y^* is the predicted label for the target node. Based on $\underline{p}$, the function CertifiedPerturbationSize computes the certified perturbation size by solving the optimization problem $M = \operatorname{argmax} \|\delta\|_0$. Their algorithm returns (y^*, M) if $\underline{p} > 0.5$ and ABSTAIN otherwise. The following proposition shows the probabilistic guarantee of our certification algorithm.

The method of Jia et al.[31] can detect the same community (for splitting attacks) or different communities (for merging attacks) when the number of edges which attacker adds or removes to a group of nodes is not greater than a certain threshold. And their method verified its effectiveness on three real-world datasets. However, their methods cannot cope with all situations, more exploration is needed in the future.

Algorithm 1: Certify

Input: f, β, x, N, α
Output: ABSTAIN or (y^*, M)
1 $m0, m_1 = SAMPLEUNDERNOISE(f, \beta, x, N, \alpha)$;
2 $y^* = \text{argmax}_{i \in \{0,1\}} m_i$;
3 $\underline{p} = B\left(\alpha; m_{y^*}, N - m_{y^*} + 1\right)$;
4 **if** $\underline{p} > 0.5$ **then**
5 | $M = CertifiedPerturbationSize(\underline{p})$;
6 | **return**(y^*, M)
7 **end**
8 **else**
9 | **return**ABSTAIN
10 **end**

6.5 Structure Based Defense

Different from the graph purification which tries to eliminate the interference, the structure based defense method aims to train a robust GNN model by punishing the model weight on the adversarial edges or nodes. These methods basically learn a kind of attention mechanism, which can distinguish the malicious edges and nodes from the clean edges and nodes, so as to reduce the impact of malicious interference on GNN training aggregation process. Zhu et al. [21] first assumed that hostile nodes may have high prediction uncertainty, because hostile nodes tend to connect the targeted node with the nodes from other communities. The work [22] showed that it is beneficial to add information of other clean graphs with similar topological distribution and node attributes to the target graph. We will introduce the following two methods in detail below: The Penalized Aggregation GNN [22] and the Robust Graph Convolutional Network [21].

6.5.1 Penalized Aggregation GNN

When the adversarial samples are fed into GNN, the aggregation function will treat the false neighbors as normal neighbors and continue to propagate their error messages to update other nodes so that the model will produce an error output. Therefore, if the messages passing through the perturbed edge are successfully filtered, the model will be hardly affected by malicious attacks.

Based on the above observation, Tang et al. [22] studied a framework Penalized Aggregation GNN (PA-GNN), aiming to improve GNN's robustness against poisoning attack by exploring a clean graph. Since the clean graphs in the real world are usually available, they created supervised knowledge to train the ability of adversarial detection by disturbing these clean graphs. Then the PA-GNN relies on a penalty aggregation mechanism, which directly limits the adversarial disturbance

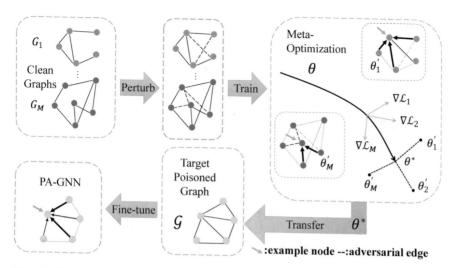

Fig. 6.6 Overall framework of PA-GNN. Thicker arrows indicate higher attention coefficients. θ^* denotes the model initialization from meta-optimization [22]

by assigning a lower attention coefficient to the perturbed edge. Besides, they also combined with the meta optimization algorithm to achieve effective defense against the target poisoning attack. Figure 6.6 shows the framework of PA-GNN. First, the clean graphs $G_1, \ldots, G_M$ are introduced to generate the perturbed edges. Then the generated disturbance is used as the supervision knowledge, and the initialization of PA-GNN is trained by meta optimization. Finally, the initialization of the target poison graph is fine tuned to get the best performance. In essence, the meta optimization is used to retain the negative effects of adversarial attacks after adapting to the target poisoned graph $\mathcal{G}$.

Specifically, inject the perturbed edges into the clean graph using the adversarial attack method at first, and then use these adversarial examples to train the ability to punish the disturbed edges. That is to say, by allocating a lower attention coefficient to the disturbed edge, only little information is transmitted to its neighbors, so that the negative impact of the adversarial disturbance can be reduced. However, because graphs have different data distribution, it is far from enough to use only supervision knowledge and punishment mechanism. Therefore, they further introduced the meta-learning algorithm, whose goal is to train a model for various learning tasks and have the capacity to meet the requirements of tasks with little or no supervised knowledge. In particular, each graph is assigned a meta optimization learning task, which not only correctly classifies the target nodes but also allocates a lower attention coefficient score to the disturbed edges of the corresponding graph.

In short, using the meta-learning algorithm, PA-GNN can resist malicious attacks when the available training data is limited. PA-GNN has been proved to be effective against three kinds of poisoning attacks based on node classification tasks: random attack, non-targeted attack and targeted attack.

6.5.2 Robust Graph Convolutional Network

Zhu et al. [21] proposed a new robust graph convolution network (RGCN) to enhance the robustness of GCNs against malicious attacks. Different from the existing defense methods, they used the Gaussian distribution as the hidden representation of nodes in the convolution layer and allocated attention weights according to the variance of nodes' neighborhood. At the same time, RGCN explicitly considers the mathematical correlation between mean and variance vector through the sampling process and regularization (Fig. 6.7).

They defined $H^{(l)} = [h_1^{(l)}, h_2^{(l)}, \ldots, h_N^{(l)}] = N(\mu_i^{(l)}, diag(\sigma_i^{(l)}))$ as the hidden representations of nodes in the lth layer for a deep learning model, where $h_1^{(l)}$ is the representation of node v_i. Applying hierarchical parameters $W^{(l)}$ and nonlinear activation functions ρ to the mean vector $\mu_i^{(l)}$ and variance vector $diag(\sigma_i^{(l)})$ of node v_i respectively, the formula of Gauss based graph convolution matrix is as follows:

$$M^{(l+1)} = \rho(\widetilde{D}^{-\frac{1}{2}} \widetilde{A} \widetilde{D}^{-\frac{1}{2}} (M^{(l)} \odot \mathscr{B}^{(l)}) W_\mu^{(l)}), \tag{6.24}$$

$$\Sigma^{(l+1)} = \rho(\widetilde{D}^{-1} \widetilde{A} \widetilde{D}^{-1} (\Sigma^{(l)} \odot \mathscr{B}^{(l)} \odot \mathscr{B}^{(l)}) W_\sigma^{(l)}), \tag{6.25}$$

where W_μ, W_σ are respectively denoted as the parameters of the mean vector and variance vector. $M^{(l)} = [\mu_1^{(l)}, \ldots, \mu_N^{(l)}]$ and $\Sigma^{(l)} = [\sigma_1^{(l)}, \ldots, \sigma_N^{(l)}]$ denote the matrix of means and variances for all nodes, respectively. $\widetilde{A} = A + I_N$, $\widetilde{D} = D + I_N$ and $\mathscr{B}^{(l)} = \exp(-\gamma \Sigma^{(l)})$. Furthermore, Since the input feature is a vector rather than a Gaussian distribution, the first layer can be represented by the full connection layer as follow:

$$M^{(1)} = \rho(H^{(0)} W_\mu^{(0)}), \quad \Sigma^{(1)} = \rho(H^{(0)} W_\sigma^{(0)}). \tag{6.26}$$

By using a Gaussian distribution as a hidden representation and assigning variance-based attention weights to the neighborhood, RGCN can reduce the effects of adversarial edges, thereby reducing the sensitivity to this adverse information. On the one hand, RGCN can improve the robustness of GCNs under non-target attack; on the other hand, the performance of RGCN is always better than the baseline regardless of the attack intensity, which indicates that its architecture can protect GCNs from various targeted attack strategies.

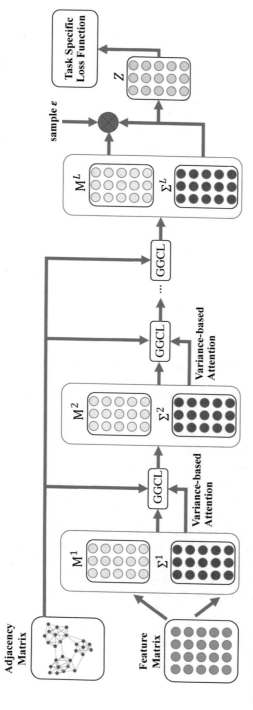

Fig. 6.7 The framework of RGCN [21]

6.6   Adversarial Detection

In many contemporary applications, it is vital to detect anomalies in graph data, such as marking fake news, exposing malicious users in social networks, preventing spam users in e-mail networks, and discovering suspicious transactions in finance.

Pezeshkpour et al. [45] automatically detected the influence of adding/deleting edges through the adversarial modification of knowledge graphs. Besides, they also studied the interpretability of knowledge graph representation. Xu et al. [23] proposed a new detection mechanism using link prediction and its variants to detect potential malicious edges. Zhang et al. [46] proposed a method to detect enemy attacks by calculating the Kullback-Leibler divergence ($K - L$ divergence) [35] average between the softmax probability of a node and its adjacent nodes. Ioannidis et al. [24] proposed a graph-based random sampling and consistency method to effectively detect abnormal nodes in large-scale graphs. Although the above detection methods can effectively find out abnormal nodes, they have strong incompatibility when facing graph-level adversarial attacks on graph classification tasks. Chen et al. [33] developed a model to detect adversarial samples on graph classification, which is based on the subgraph network (SGN). In the following, we will respectively introduce the adversarial detection methods on node classification and graph classification tasks.

6.6.1   Adversarial Detection on Node Classification

Zhang et al. [46] studied the influence of the recently proposed attack method on the GCN model and developed a detection method to detect the adversarial nodes. Specifically, they first studied the random attack and Nettack [12] on graph deep learning model and found that these attack methods interfere with the graph structure. Secondly, they also studied the statistical differences between undisturbed and disturbed graphs and further proved that topological perturbation is more important than characteristic perturbation. Therefore, this work assumes that there is no characteristic disturbance.

Essentially, the prediction logic used for classification is similar to the node embedding. Since the Nettack uses the logits of the node v_i, they expected Nettack to produce differences between the first-order neighbor information of v_i and that of the neighbors of v_i. They measured this difference by averaging the $K - L$ divergence difference between the softmax probabilities of v_i and its neighbors:

$$\text{pr}_1(i) = \frac{1}{|ne(i)|} \sum_{j \in ne(i)} D_{\text{KL}}(p_i \| p_j), \tag{6.27}$$

where p_i is the softmax probabilities of GCN output for the ith node. In addition, $K - L$ divergence between the softmax probabilities of neighborhood pairs is used to calculate the second-order proximity information:

$$\text{pr}_2(i) = \frac{1}{|ne(i)|(|ne(i)| - 1)} \sum_{j \in ne(i)} \sum_{k \in ne(i)} D_{\text{KL}}(p_j \| p_k). \tag{6.28}$$

The authors defined straightforward detection tests by setting thresholds τ_1 and τ_2 for pr_1 and pr_2, respectively. Given a node that may be interfered, if pr_1 exceeds τ_1 or pr_2 exceeds τ_2, the node is marked as the adversarial. They further utilized Neyman-Pearson lemma [47] to set the different detection thresholds for each dataset.

The null hypothesis distribution is modeled as a normal distribution. Moreover, the maximum likelihood method is used to fit the Gaussian distribution of the undisturbed training data. Matching the tail probability with the false positive rate of a specific target is to find the appropriate detection threshold. In particular, the threshold is calculated by the inverse Cumulative Distribution Function (CDF).

6.6.2 Adversarial Detection on Graph Classification

For graph classification, we also propose an adversarial detection model based on SGN. The overall framework is shown in Fig. 6.8. In particular, our method detects adversarial samples by evaluating the original input features and the transformed features of SGNs. If the difference between the predicted value of the original input sample and the input sample after subgraph transformation exceeds a certain threshold, the discriminator will recognize the input as malicious samples, otherwise it will be clean samples. Note that the detailed description of SGN is provided in Chap. 3, and we will not introduce it here.

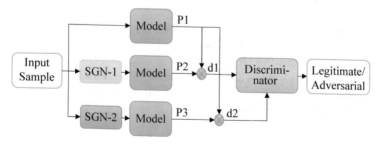

Fig. 6.8 The framework for SGN based adversarial detection [33]

6.6.2.1 SGN Based Adversarial Detection

The basic idea behind our SGN based adversarial detection algorithm is to find out the key points to distinguish the adversarial samples from the clean ones by comparing the sample characteristics before and after the mapping of subgraph networks. In other words, The prediction and reconstructed version of the model for a legitimate example should be similar. Conversely, if the original example and the reconstructed example produce very different predictions, the input is likely to be adversarial. SGN is able to make accurate model predictions for many adversarial examples, while the prediction accuracy for legal examples is hardly reduced.

The prediction vector generated by graph classifier usually represents the probability distribution that the input samples belong to each possible class. Therefore, the original prediction value of the model is compared with the predicted value of the reconstructed sample, in other words, the corresponding two probability distribution vectors are compared. In this work, we choose the L_1 norm of the original prediction vector and the reconstructed prediction vector as the difference measure between the adversarial sample and the clean sample:

$$d^{(x,x_{sgn})} = ||P(x) - P(x_{sgn})||_1. \tag{6.29}$$

We can also try L_2 norm and $K - L$ divergence to measure the difference of probability distribution. $p(x)$ is the output vector generated by the input samples in the softmax layer of the graph classifier. The higher the d is, the more significant the difference between the original prediction and the SGN based reconstruction prediction. In fact, the value of d is expected to be as low as possible under the input of legal samples and as high as possible under the input of adversarial samples. In this way, it is easy to find the most appropriate and optimal threshold to distinguish between adversary and legal samples.

6.6.2.2 Joint Adversarial Detection

In the real world, even if we can choose the appropriate effective SGN based adversarial detection model for a specific type of attack, we can't predict which attack strategies the attacker will use to pollute the sample. In order to meet this challenge, we use multi-order SGN to construct a joint adversarial detection model. More specifically, we use the maximum distance between $d^{(x,x_{sgn1})}$ and $d^{(x,x_{sgn2})}$ as a measure of the adversarial sample:

$$d^{joint} = max(d^{(x,x_{sgn1})}, d^{(x,x_{sgn2})}, \dots). \tag{6.30}$$

At present, joint adversarial sample detection can effectively detect the overwhelming majority of adversarial samples generated for graph classification task. In the actual joint adversarial samples detection model, SGN$^{(1)}$ and SGN$^{(2)}$ are mainly used to transform the sample features, since compared with the high-order SGNs, they are simpler to construct, easier to implement, and has lower time complexity. However, the introduction of Max operator often leads to the most destructive feature reconstruction of legal input samples, which will greatly improve the False Positive Rates (FPR).

6.7   Summary of Defenses

In this chapter, we will compare the various defense methods and summarize the advantages and disadvantages.

Recently, the relatively mature defense methods to improve the robustness of GNNs mainly include adversarial training, graph purification and structure based defense methods, while robustness certification and adversarial detection are still in vigorous development. Compared with the other defense methods, the structure based defense method mostly uses attention mechanism to optimize and improve the robustness of model. In addition, the biggest difference between the adversarial detection and other defense methods is that it does not directly improve the robustness of GNNs, but attempts to detect the disturbed sample.

As shown in Table 6.1, most of the existing defense work [20–22, 28–30, 32, 46] focus on the node classification task, and the research of other important graph data mining tasks is still very scarce. In detail, only a few papers have done research on model robustness of community detection [28, 31] and graph classification [33] recently. Therefore, it is of great significance and value to improve the robustness of the model for various tasks or to transfer the existing defense methods to other tasks.

Time complexity is of high significance in real applications. Therefore, how to limit the training cost and meanwhile improve the robustness of the model is also a valuable research direction. However, the current defense methods rarely consider the space-time efficiency of their algorithms. Some studies [32, 48] have used different dimensionality reduction methods to reduce costs and achieve higher efficiency, but their experiments did not involve large-scale graphs. Another work [49] is to discretize the regularized adjacency matrix in the training process, which can effectively improve the efficiency.

6.8   Experiment and Analyze

6.8.1   Adversarial Training

In order to verify the effectiveness of SAT algorithm, we test it in two common graph data mining tasks: node classification and community detection. The experiment uses three networks: PolBlogs, Cora and Citeseer. We adopt FGA [11], and Nettack

Table 6.1 Categorization of representative defend methods

Method	Type	Target task	Target model	Baseline	Metric	Dataset
RGCN[21]	Structure based	Node classification	GCN	GCN,GAT	Accuracy	Citeseer, Cora, Pubmed
PA-GNN[22]	Structure based	Node classification	GNN	GCN,GAT, PreProcess, RGCN,VPN	Accuracy	Pubmed, Reddit, Yelp-Small, Yelp-Large
GAT,GATv[34]	Adversarial training	Node classification		DeepWalk, GCN, GraphSGAN, …	Accuracy	Citeseer, Cora, NELL
Global-AT, Target-AT, SD,SCEL[28]	Adversarial training, Community Detection	Node classification	GNN	AT, GraphDefense	ADR, ACD	Citeseer, Cora, PolBlogs
GCN-Jaccard[29]	Graph Purification	Node classification	GCN	GCN	Accuracy, Classification margin	Citeseer, Cora-ML, PolBlogs
GCN-SVD[30]	Graph Purification	Node classification	GCN, t-PINE	GCN, t-PINE	Accuracy	Citeseer, Cora-ML, PolBlogs
[20]	Robustness Certification	Node classification	Pagerank	Pagerank	F1 score, Accuracy	Citeseer, Cora-ML
[31]	Robustness Certification	Community Detection	Louvain's method	Louvain's method	certified accuracy	Email, DBLP, Amaz
[32]	Robustness Certification	Node classification	GCN	GCN	Accuracy	Citeseer, Cora-ML, Pubmed
[33]	Adversarial Detection	graph classification	DNN	DNN	F1 score, SAR,FAR, FP,AUC	MUTAG, DHFR, BZR
[46]	Adversarial Detection	Node classification	GCN,GAT	-	Accuracy, Classification margin, ROC,AUC	Citeseer, Cora, PolBlogs
GraphSAC[24]	Adversarial Detection	Anomaly Detection	Anomaly model	GAE Degree, Cut ratio,…	AUC	Citeseer, PolBlogs, Cora, Pubmed

[39] as the adversarial attacks, where FGA is a gradient-based attack method, and Nettack is a score-based attack method. Both of these adversarial attack methods achieve good attack success rates on GNN models.

We use the following metrics to measure the defense effectiveness on the attacked nodes, where the nodes are classified correctly before attacked in the test set.

- **Average Defense Rate (ADR).** ADR indicates the relative difference between the ASR of attack on GCN with and without defense. It can be calculated as:

$$ADR = ASR_{atk} - ASR_{def}, \tag{6.31}$$

where ASR_{atk} is the ASR without defense methods, and ASR_{def} is the ASR with defense methods. The higher ADR corresponds to the better defense effect.
- **Average Confidence Different (ACD).** The average confidence difference, i.e, the average difference between the confidences of the nodes in n_{suc} before and after attack, which is defined as follow:

$$ACD = \frac{1}{n_{suc}} \sum_{t \in N_s} CD_i(\hat{A}_t) - CD_i(A), \tag{6.32}$$

$$CD_i(A) = \max_{c \neq y} Y'_{i,c}(A) - Y'_{i,y}(A), \tag{6.33}$$

where $\hat{A}_t$ is the adversarial network of target node t, y is the real label of target node t, Y' is the output of the model. The lower ACD corresponds to the better defense effect.

First, we use GCN as the basic model and verify the effectiveness of the defense method in the node classification task. The ADR and ACD of the four defense methods are reported, respectively. Figure 6.9a and b shows the average ACR% (a) and ACD (b) of the two attack methods in each dataset. It can be seen that Target-AT has the highest ADR% and the lowest ACD in most cases, so it is the optimal defense strategy. The defense effects of Global-AT and SCEL defense strategies are not much different, while SD behaves the worst among all defense methods.

Besides, we combine two adversarial training strategies to extend four combined defense mechanisms. G-SD stands for the combined defense of Global-AT and SD technology, T-SCEL stands for the combined defense of target-at and SCEL technology, and other combined defenses are the same. Figure 6.9c and d reports the defense results of these four combined defenses and an advanced defense mechanism, GraphDefense. It can be seen that these combined defense mechanisms combining adversarial training and smoothing strategy can improve the defense effect. We believe that this is because the two defenses have complementary effects. At the same time, for the four combination strategies, the T-SCEL defense mechanism has the highest ADR and the lowest ACD. SCEL has more obvious improvement than SD in these two adversarial training methods. Finally, almost all

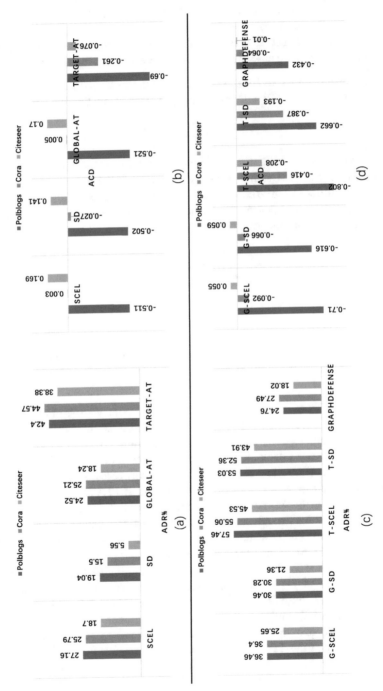

Fig. 6.9 The defense results of the four defense methods and the four combined defense methods in the node classification task

combined defense methods are superior to GraphDefense, which fully demonstrates the effectiveness of the combined defense.

Next, we verify the defensive effect of several types of defense methods on community detection tasks. The datasets we used are PloBook and Dolphins. The PloBook is the joint purchase dataset of co-purchasing of books about US politics sold by the online bookseller and the Dolphins is the social network among 62 dolphins in New Zealand. We also use the FGA and Nettack attack methods to attack the three community detection algorithms: DeepWalk, Node2Vec, and Louvain respectively. Furthermore, we show the defense effect of eight defense methods: four independent defense and four combined defense, and compare them with GraphDefense as shown in Fig. 6.10. Note that we just show the average ADR of three community detections algorithm. It can be found that independent defense usually cannot achieve good performance. At the same time, T-SCEL can obtain the optimal defense effect in most cases, which is consistent with the results in node classification.

6.8.2 Adversarial Detection

We use the following three most commonly used datasets for graphs classification adversary detection: MUTAG, DHFR and BZR. Furthermore, random attack and gradient attack are used to attack graph-classification model. Each dataset is randomly divided into two groups: one is used to train the adversarial detection model, and the other is used to verify the detection results. We use the original training dataset to generate the same number of adversarial samples, and use these two kinds of sample sets to train the detection model. After the training, the test datasets (half clean samples, half adversarial samples) were input to testify the accuracy of the joint detection model.

In essence, the training phase of our detector is to select an optimal threshold to distinguish clean and adversarial samples. For a legitimate example, the predicted and reconstructed versions of the model should be similar. Conversely, if the original and reconstructed examples produce very different predictions, the input is hostile. Figure 6.11 visually demonstrates this by comparing the predicted d values between the original and the attacked samples. Because the expected distribution of samples is unbalanced, and most of them are benign, the detector with high precision but high false alarm rate is useless for many security sensitive tasks. Therefore, we need to select the false positive rate below 7% as the target threshold, in other words, select the threshold value of a legal sample not exceeding 7%. After the training, we use the selected threshold to test, and measure the Detection Rate of Successful Adversarial Sample (SADR) and the Detection Rate of Failed Adversarial Sample (FADR), respectively.

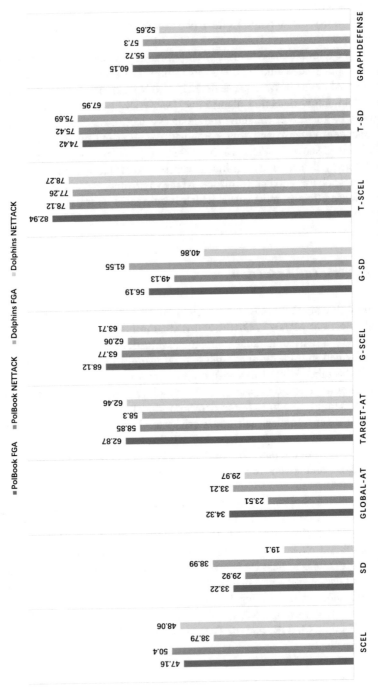

Fig. 6.10 The average defense results of the four defense methods and the four combined defense methods in the community detection task

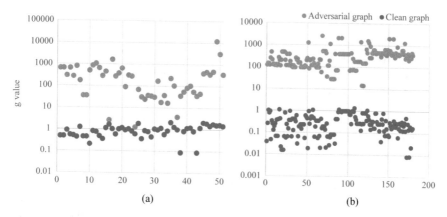

Fig. 6.11 The d value measuring clean and adversarial input. (a) MUTAG examples. (b) BZR examples

Table 6.2 Detection results for different kinds of the adversarial detectors on different datasets

Dataset	Detector	Random attack			Grad attack		
		SADR	FADR	ROC-AUC	SADR	FADR	ROC-AUC
MUTAG	SGN-1 detector	86.30%	12.13%	89.10%	88.53%	12.52%	90.92%
	Joint detector	**93.65%**	**13.77%**	**95.32%**	**95.28%**	**14.07%**	**97.25%**
DHFR	SGN-1 detector	**82.13%**	**10.65%**	**84.23%**	85.83%	11.53%	87.26%
	Joint detector	82.13%	10.65%	84.23%	**86.20%**	**11.82%**	**88.52%**
BZR	SGN-1 detector	89.64%	16.37%	92.36%	87.33%	13.01%	91.42%
	Joint detector	**96.58%**	**18.57%**	**98.22%**	**94.25%**	**13.57%**	**96.74%**

the bold value indicates the relative optimal detector with the best detection index under the corresponding data set and attack method

As shown in Table 6.2, it can be seen that the overall detection performance of the joint detector is generally better than that of the SGN$^{(1)}$ detector, because it is actually a comprehensive detector of the first-order and second-order SGNs. In some cases, SGN$^{(1)}$ detector can directly meet the requirements of gradient attack on DHFR dataset and reinforcement learning on BZR dataset. But in most other cases, it is necessary to establish a joint detection system because the subgraph networks of different orders have different matching degrees for different attack methods and different dataset structures, and model operators are unlikely to know in advance what attacks will be used. Secondly, the detection rate of successful adversarial samples in all datasets is above 80%. Although SADR is high, FADR value is very low. In fact, FADR does not affect the normal detection work, because the failed attack samples will not interfere with the task of graph classification. In addition, the ROC-AUC score of our model is high, proving the effectiveness of our detection model.

6.9 Conclusion

In this chapter, we conducted a comprehensive review of adversarial learning, including defense strategies and corresponding evaluation indicators. Specifically, we introduced the latest developments in this field and the arms race between the offensive and the defensive. In addition, we classified the defense methods reasonably, and give a unified problem expression to make it clear and easy to understand. In various typical scenarios, we also summarize and discuss the main contributions and limitations of existing defense methods, as well as the open issues worthy of exploration in this field. Our work also covers most of the relevant assessment indicators in the field of graph adversarial learning to provide a better understanding of these methods. We proposed two adversarial training strategies and two smoothing strategies in order to defend against malicious attacks of node classification and community detection. Moreover, we also addressed the challenge of the robustness of graph classification, especially to find adversarial samples. Through extensive experiments, the effectiveness of adversarial training and adversarial detection method is proved. In the future, we will focus on other more effective and less time-consuming defense methods to defend against increasingly complex and diverse attack methods.

References

1. Ahmed, A., Al-Masri, N., Abu Sultan, Y.S., Akkila, A.N., Almasri, A., Mahmoud, A.Y., Zaqout, I.S. and Abu-Naser, S.S. Knowledge-based systems survey International Journal of Academic Engineering Research (IJAER) 3(7), 1–22 (2019)
2. Pei, Y., Chakraborty, N., Sycara, K.: Nonnegative matrix tri-factorization with graph regularization for community detection in social networks. In: Twenty-Fourth International Joint Conference on Artificial Intelligence (2015)
3. Wang, J., Huang, P., Zhao, H., Zhang, Z., Zhao, B., Lee, D.L.: Billion-scale commodity embedding for e-commerce recommendation in alibaba. In: Proceedings of the 24th ACM SIGKDD International Conference on Knowledge Discovery & Data Mining, pp. 839–848 (2018)
4. Xie, F., Chen, L., Ye, Y., Zheng, Z., Lin, X.: Factorization machine based service recommendation on heterogeneous information networks. In: 2018 IEEE International Conference on Web Services (ICWS), pp. 115–122. IEEE, Piscataway (2018)
5. Kipf, T.N., Welling, M.: Semi-supervised classification with graph convolutional networks. Preprint. arXiv:1609.02907 (2016)
6. Hamilton, W., Ying, Z., Leskovec, J.: Inductive representation learning on large graphs. In: Advances in Neural Information Processing Systems, pp. 1024–1034 (2017)
7. Bruna, J., Zaremba, W., Szlam, A., LeCun, Y.: Spectral networks and locally connected networks on graphs. Preprint. arXiv:1312.6203 (2013)
8. Defferrard, M., Bresson, X., Vandergheynst, P.: Convolutional neural networks on graphs with fast localized spectral filtering. In: Advances in Neural Information Processing Systems, pp. 3844–3852 (2016)

9. Ying, Z., You, J., Morris, C., Ren, X., Hamilton, W., Leskovec, J.: Hierarchical graph representation learning with differentiable pooling. In: Advances in Neural Information Processing Systems, pp. 4800–4810 (2018)
10. Zhang, M., Cui, Z., Neumann, M., Chen, Y.: An end-to-end deep learning architecture for graph classification. In: Thirty-Second AAAI Conference on Artificial Intelligence (2018)
11. Chen, J., Wu, Y., Xu, X., Chen, Y., Zheng, H., Xuan, Q.: Fast gradient attack on network embedding. Preprint. arXiv:1809.02797 (2018)
12. Zügner, D., Akbarnejad, A., Günnemann, S.: Adversarial attacks on neural networks for graph data. In: Proceedings of the 24th ACM SIGKDD International Conference on Knowledge Discovery & Data Mining, pp. 2847–2856 (2018)
13. Yu, S., Zhao, M., Fu, C., Zheng, J., Huang, H., Shu, X., Xuan, Q., Chen, G.: Target defense against link-prediction-based attacks via evolutionary perturbations. IEEE Trans. Knowl. Data Eng. 33(2), 754–767 (2021)
14. Zügner, D., Günnemann, S.: Adversarial attacks on graph neural networks via meta learning. Preprint. arXiv:1902.08412 (2019)
15. Chen, J., Zhang, J., Chen, Z., Du, M., Li, F., Xuan, Q.: Time-aware gradient attack on dynamic network link prediction. Preprint. arXiv:1911.10561 (2019)
16. Dai, H., Li, H., Tian, T., Huang, X., Wang, L., Zhu, J., Song, L.: Adversarial attack on graph structured data. Preprint. arXiv:1806.02371 (2018)
17. Tang, H., Ma, G., Chen, Y., Guo, L., Wang, W., Zeng, B., Zhan, L.: Adversarial attack on hierarchical graph pooling neural networks. Preprint. arXiv:2005.11560 (2020)
18. Ma, Y., Wang, S., Derr, T., Wu, L., Tang, J.: Attacking graph convolutional networks via rewiring. Preprint. arXiv:1906.03750 (2019)
19. Zhang, Z., Jia, J., Wang, B., Gong, N.Z.: Backdoor attacks to graph neural networks. Preprint. arXiv:2006.11165 (2020)
20. Bojchevski, A., Günnemann, S.: Certifiable robustness to graph perturbations. In: Advances in Neural Information Processing Systems, pp. 8319–8330 (2019)
21. Zhu, D., Zhang, Z., Cui, P., Zhu, W.: Robust graph convolutional networks against adversarial attacks. In: Proceedings of the 25th ACM SIGKDD International Conference on Knowledge Discovery & Data Mining, pp. 1399–1407 (2019)
22. Tang, X., Li, Y., Sun, Y., Yao, H., Mitra, P. and Wang, S.: Transferring robustness for graph neural network against poisoning attacks. In: Proceedings of the 13th International Conference on Web Search and Data Mining, pp. 600–608 (2020)
23. Xu, X., Yu, Y., Li, B., Song, L., Liu, C., Gunter, C.: Characterizing malicious edges targeting on graph neural networks (2019) https://openreview.net/forum?id=HJxdAoCcYX
24. Ioannidis, V.N., Berberidis, D., Giannakis, G.B.: Graphsac: Detecting anomalies in large-scale graphs. Preprint. arXiv:1910.09589 (2019)
25. Goodfellow, I.J., Shlens, J., Szegedy, C.: Explaining and harnessing adversarial examples. Preprint. arXiv:1412.6572 (2014)
26. Kurakin, A., Goodfellow, I., Bengio, S.: Adversarial machine learning at scale. Preprint. arXiv:1611.01236 (2016)
27. Miyato, T., Dai, A.M., Goodfellow, I.: Adversarial training methods for semi-supervised text classification. Preprint. arXiv:1605.07725 (2016)
28. Chen, J., Lin, X., Xiong, H., Wu, Y., Zheng, H., Xuan, Q.: Smoothing adversarial training for GNN IEEE Trans. Comput. Soc. Syst. 1–12 (2020)
29. Wu, H., Wang, C., Tyshetskiy, Y., Docherty, A., Lu, K., Zhu, L.: Adversarial examples on graph data: Deep insights into attack and defense. Preprint. arXiv:1903.01610 (2019)
30. Entezari, N., Al-Sayouri, S.A., Darvishzadeh, A., Papalexakis, E.E.: All you need is low (rank) defending against adversarial attacks on graphs. In: Proceedings of the 13th International Conference on Web Search and Data Mining, pp. 169–177 (2020)
31. Jia, J., Wang, B., Cao, X., Gong, N.Z.: Certified robustness of community detection against adversarial structural perturbation via randomized smoothing. In: Proceedings of The Web Conference 2020, pp. 2718–2724 (2020)

32. Zügner, D., Günnemann, S.: Certifiable robustness and robust training for graph convolutional networks. In: Proceedings of the 25th ACM SIGKDD International Conference on Knowledge Discovery & Data Mining, pp. 246–256 (2019)

33. Chen, J., Xu, H., Wang, J., Xuan, Q., Zhang, X.: Adversarial detection on graph structured data. In: Proceedings of the 2020 Workshop on Privacy-Preserving Machine Learning in Practice, pp. 37–41 (2020)

34. Feng, F., He, X., Tang, J., Chua, T.S.: Graph adversarial training: Dynamically regularizing based on graph structure. IEEE Trans. Knowl. Data Eng. **33**(6), 2493–2504 (2021)

35. Van Erven, T., Harremos, P.: Rényi divergence and kullback-leibler divergence. IEEE Trans. Inf. Theory **60**(7), 3797–3820 (2014)

36. Miyato, T., Maeda, S.I., Koyama, M. and Ishii, S.: Virtual adversarial training: a regularization method for supervised and semi-supervised learning. IEEE Trans. Pattern Anal. Mach. Intell. **41**(8), 1979–1993 (2018)

37. Hinton, G., Vinyals, O., Dean, J.: Distilling the knowledge in a neural network. Preprint. arXiv:1503.02531 (2015)

38. Szegedy, C., Vanhoucke, V., Ioffe, S., Shlens, J., Wojna, Z.: Rethinking the inception architecture for computer vision. In: Proceedings of the IEEE Conference on Computer Vision and Pattern Recognition, pp. 2818–2826 (2016)

39. Wong, E., Kolter, Z.: Provable defenses against adversarial examples via the convex outer adversarial polytope. In: International Conference on Machine Learning, pp. 5286–5295. PMLR (2018)

40. Chen, J., Chen, L., Chen, Y., Zhao, M., Yu, S., Xuan, Q., Yang, X.: GA-based Q-attack on community detection. IEEE Trans. Comput. Soc. Syst. **6**(3), 491–503 (2019)

41. Chen, Y., Nadji, Y., Kountouras, A., Monrose, F., Perdisci, R., Antonakakis, M., Vasiloglou, N.: Practical attacks against graph-based clustering. In: Proceedings of the 2017 ACM SIGSAC Conference on Computer and Communications Security, pp. 1125–1142 (2017)

42. Fionda, V., Pirro, G.: Community deception or: How to stop fearing community detection algorithms. IEEE Trans. Knowl. Data Eng. **30**(4), 660–673 (2017)

43. Nagaraja, S.: The impact of unlinkability on adversarial community detection: effects and countermeasures. In: International Symposium on Privacy Enhancing Technologies Symposium, pp. 253–272. Springer, Berlin (2010)

44. Waniek, M., Michalak, T.P., Wooldridge, M.J., Rahwan, T.: Hiding individuals and communities in a social network. Nat. Hum. Behav. **2**(2), 139–147 (2018)

45. Pezeshkpour, P., Tian, Y., Singh, S.: Investigating robustness and interpretability of link prediction via adversarial modifications. Preprint. arXiv:1905.00563 (2019)

46. Zhang, Y., Khan, S., Coates, M.: Comparing and detecting adversarial attacks for graph deep learning. In: Proceedings of Representation Learning on Graphs and Manifolds Workshop, International Conference Learning Representations, New Orleans, LA (2019)

47. Series, A: Containing papers of a mathematical or physical character. Vol. CLXXIX (1888)

48. Wang, S., Chen, Z., Ni, J., Yu, X., Li, Z., Chen, H., Yu, P.S.: Adversarial defense framework for graph neural network. Preprint. arXiv:1905.03679 (2019)

49. Wang, X., Liu, X., Hsieh, C.-J.: Graphdefense: Towards robust graph convolutional networks. Preprint. arXiv:1911.04429 (2019)

Chapter 7
Understanding Ethereum Transactions via Network Approach

Yunyi Xie, Jiajun Zhou, Jinhuan Wang, Jian Zhang, Yunxuan Sheng,
Jiajing Wu, and Qi Xuan

Abstract Ethereum, which is one of the largest public blockchain-based platforms, provides an unprecedented opportunity for data mining. However, the analysis of Ethereum transaction records remains under-explored. In this chapter, we model Ethereum transaction records as a complex network and further study the problem of phishing detection and transaction tracking via node classification and link prediction, respectively, which provides a deeper understanding of Ethereum transactions from a network perspective. Specifically, we introduce time-series snapshot network (TSSN) to model Ethereum transaction records as a spatial-temporal network and present temporal biased walk (TBW) to effectively embed accounts via their transaction records, which integrates temporal and structural information of the proposed network. Furthermore, we provide a detailed and systematic analysis of various graph embedding models and compare our proposed method with these embedding technologies on realistic Ethereum transaction records. Experimental results demonstrate the superiority of our TBW in learning more informative representations, and is essential for Ethereum network analysis.

Y. Xie · J. Zhou · J. Wang · J. Zhang · Q. Xuan (✉)
Institute of Cyberspace Security, Zhejiang University of Technology, Hangzhou, China

College of Information Engineering, Zhejiang University of Technology, Hangzhou, China
e-mail: xuanqi@zjut.edu.cn

Y. Sheng
School of Computer Science, Faculty of Science and Engineering, University of Nottingham
Ningbo China, Ningbo, China

J. Wu
School of Computer Science and Engineering, Sun Yat-sen University, Guangzhou, China

© The Author(s), under exclusive license to Springer Nature Singapore Pte Ltd. 2021   155
Q. Xuan et al. (eds.), *Graph Data Mining*, Big Data Management,
https://doi.org/10.1007/978-981-16-2609-8_7

7.1 Introduction

Blockchain is an open distributed ledger, which can effectively, verifiably, and permanently record transactions among peers [1]. As the largest public blockchain-based platform enabling smart contracts, Ethereum provides available transaction records that contain rich historical information and can be used to study Ethereum matters. However, with the rapid development of blockchain technology, Ethereum has become a hotbed of various cybercrimes [2]. Taking advantage of the anonymity of blockchain, criminals attempt to evade supervision and engage in illegal activities by injecting funds into the blockchain system. It's reported that Ethereum has suffered from a variety of scams, such as hacks, phishing, and Ponzi schemes [3], showing that cybercrimes have become a critical issue in Ethereum. In order to create a favorable investment environment and preserve the sustainable development of blockchain-based systems, it's imperative to formulate effective supervision.

In this chapter, we focus on the solution of two illegal activities on Ethereum, i.e., phishing detection and transaction tracking. Recent years have witnessed the rise of online business, phishing scams emerge as the main threat to trading security, and thus an effective method for detection and prevention of phishing scams is urgently needed. Transaction tracking, which focuses on maintaining transaction security, is capable of identifying fraud groups, tracing capital flows, retrieving stolen money, and improving the regulatory system. Besides, transaction tracking helps ordinary investors or cryptocurrency companies check whether certain funds or transactions are associated with illegal entities or contaminated by suspicious paths. In a word, it is necessary to formulate effective regulatory measures to prevent crime. The issues of phishing detection and transaction tracking have been widely discussed and many methods have been proposed. However, compared with traditional scenarios, illegal activities on Ethereum behave very differently. Traditional illegal activities generally rely on phishing emails and websites to obtain users' sensitive information, which leads to existing methods focusing on how to detect emails or websites containing phishing fraud information [4], so existing methods cannot be directly applied to illegal activities problem on Ethereum.

Fortunately, all the historical transaction records of Ethereum are publicly accessible, which may help to address the illegal problem. Here, we model Ethereum transaction records as a transaction network for further understanding and study phishing detection and transaction tracking on Ethereum from a network perspective. Generally speaking, a network (or graph) is adopted as a standard representation of data, which occurs in various real-world scenarios. More recently, graph embedding, an effective method to represent node features in a low dimensional space for network analysis, has been widely applied to graph analytics problem. In this way, the graph structural information and graph properties are maximumly preserved and thus can benefit a lot of useful downstream machine learning tasks such as node classification [5], link prediction [6], community detection [7], etc. Intuitively, the huge Ethereum transaction records can be modeled as a transaction network, where a node represents an account and an edge connecting two nodes corresponds to the existence of at least one transaction between them. Specifically, the problem of phishing detection on Ethereum can be modeled as a binary classification problem, while tracing and predicting transactions on the transaction network is a link prediction task.

The study of transaction network has recently attracted considerable attention from different applications like graph analysis [8], price prediction [9], and anti market manipulation [10]. By building a monthly trading network, Liang et al. [11] tracked the dynamics of three representative cryptocurrencies over time, namely Bitcoin, Ethereum, and Namecoin. Wu et al. [12] proposed the concept of *Attributed Temporal Heterogeneous motifs* and further addressed the issue of mixing detection using a detection model. More recently, they presented a graph embedding model named *trans2vec* for transaction networks, which incorporate the transaction amount values and timestamps. It is worth noting that the model makes contributes to phishing detection [13, 14], and transaction tracking [15], and can be applied to other similar scenarios on transaction networks. Our method, however, is different from *trans2vec*. We define a temporal-spatial network named time-series snapshot network (TSSN) to model the Ethereum transaction records and make the successive snapshots connect to reduce temporal loss. Furthermore, we introduce temporal biased walk (TBW) to learn the representations of accounts. For each account, a unique searching strategy is adopted. The searching strategy depends on the number of transactions, structural transition probability, and temporal transition probability. Various experiments conducted on real-world Ethereum datasets demonstrate that our approach can efficiently learn informative account representation, and solve the problem of phishing detection and transaction tracking.

The rest of this chapter is organized as follows. In Sect. 7.2, we introduce Ethereum transaction dataset and construct networks needed for subsequent experiments. In Sect. 7.3, we summarize the related work on graph embedding. In Sect. 7.4 we give the basic definitions and present our method. In Sect. 7.5, we conduct extensive experiments on real-world Ethereum datasets and compare our method with several graph embedding techniques. Finally, we conclude this chapter in Sect. 7.6.

7.2 Ethereum Transaction Dataset

As the largest public blockchain-based platform, Ethereum's [16] transaction records are completely public, which brings unprecedented opportunities for transaction network analysis. Ethereum introduces *account* which is an address allocating storage space for recording account balances, transactions, codes, etc. [17]. Accounts can be divided into two categories, i.e., external owned accounts (EOAs) and smart contracts [18]. The main difference between them is that smart contracts contain executable code files. Both smart contracts and EOAs can participate in transactions, whereas EOA participates in transactions just like general bank

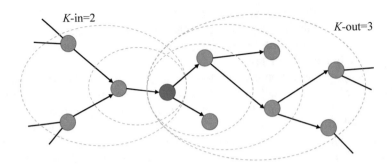

Fig. 7.1 A schematic depiction of K-order subgraph. The central node and its local neighbors are marked in blue and orange, respectively

accounts. In this chapter, we focus on the transactions among EOAs. Note that those with zero amount value are removed for that we do not consider such records here.

Thanks to the openness of blockchain, researchers can independently access Ethereum transaction records and easily obtain the historical transaction data of the target account through the API of Etherscan (https://etherscan.io). The size of the total transaction records is extremely large and the majority of accounts in Ethereum transactions are not always active, which may result in extremely large and sparse Ethereum transaction networks when processing the whole Ethereum transaction records [19]. Therefore, we select several target accounts and obtain their transactions from Ethereum transaction records to make subgraphs for subsequent experiments. As shown in Fig. 7.1, we randomly sample a centered account to obtain its local structure information and then extract K-order subgraph [20]. K-in and K-out are two parameters to control the depth of sampling inward and outward from the center, respectively. The visualization of real Ethereum transactions constructed by the K-order subgraph is shown in Fig. 7.2.

We study the problem of phishing detection and transaction tracking via machine learning tasks: link prediction and node classification. In the following experiments, we collect four subgraphs with different sizes from the whole Ethereum transaction records, three for link prediction, and one for node classification. We randomly select different center accounts and collect three subgraphs with K-in = 1, K-out = 3 for link prediction task. As for node classification task, we assume that the previous account of the phishing account may be a victim, and the next three accounts may be the bridge accounts with money laundering behavior. Therefore, we collect subgraphs with K-in = 1, K-out = 3 for each of 445 labeled phishing accounts and 445 unlabeled accounts (treated as non-phishing accounts), and then splice them into a large-scale network. The basic topological features of these networks are listed in Table 7.1.

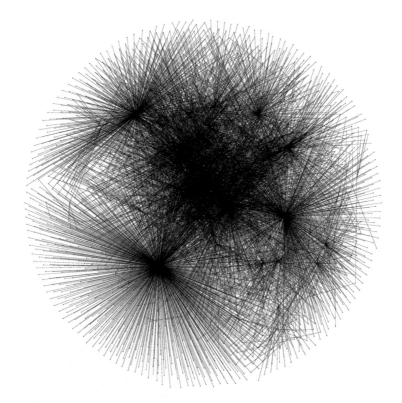

Fig. 7.2 Visualization of the network collected from real Ethereum transactions constructed by the K-order subgraph

Table 7.1 Basic topological features of networks. $|V|$ and $|E|$ are the numbers of nodes and edges, respectively, $\langle K \rangle$ is the average degree, and $\langle C \rangle$ is the clustering coefficient

	$|V|$	$|E|$	$\langle K \rangle$	$\langle C \rangle$
EthereumG1	2100	6995	6.662	0.211
EthereumG2	5762	9098	3.158	0.112
EthereumG3	10,269	28,431	5.537	0.147
EthereumG	86,622	104,322	2.409	0.038

7.3 Graph Embedding Techniques

Generally speaking, graph embedding methods convert nodes in the graph into low dimensional vector representations, in which the structural information and topology properties of nodes are preserved as much as possible [21, 22]. In this section, we provide a detailed and systematic review of various graph embedding methods and divide these embedding methods into four categories: (1) factorization methods, (2) random walk based methods, (3) deep learning, and (4) other miscellaneous strategies. Below we explain the characteristics of these categories and provide a summary of some representative methods for each category.

7.3.1 Factorization Based Methods

Factorization based graph embedding represents the connections between nodes in the form of a matrix and obtains the node embedding by factorizing this matrix. There are several kinds of matrices used to represent the connections between nodes, such as node transition probability matrix, node adjacency matrix, Laplacian matrix, and so on. Factorization based methods vary with the matrix properties. One can use eigenvalue decomposition for a positive semidefinite matrix such as the Laplacian matrix. But for unstructured matrices, one can obtain the embedding in linear time by employing gradient descent methods.

- **Graph Factorization (GF)** [23] is the first method to obtain a graph embedding in $O(|E|)$ time. To obtain the embedding, GF factorizes the graph's adjacency matrix by minimizing the loss function: $f(Y, \lambda) = \frac{1}{2} \sum_{(i,j) \in E} (W_{ij} - \langle Y_i, Y_j \rangle)^2 + \frac{\lambda}{2} \sum_i \|Y_i\|^2$, where W is the weight matrix, Y is the reconstructed embedding matrix, and λ is a regularization coefficient. The summation is the edges that can be observed rather than all possible edges, which is an approximation for scalability and may introduce noise. The adjacency matrix is generally not positive semidefinit, thus the minimum of loss function is greater than 0 even if the dimensionality of embedding is $|V|$.
- **HOPE** [24] preserves higher order proximity by minimizing $\|S - Y_s Y_t^T\|_F^2$, where S is the similarity matrix, e.g., katz index, rooted page rank, common neighbors, and adamic-adar score. The similarity measure can be represented as $S = M_g^{-1} M_l$, where M_g and M_l are matrix polynomials, which correspond to the network information of global and local respectively, and both of them are sparse. This enables HOPE to get the embedding efficiently by generalized singular value decomposition (SVD) [25].

7.3.2 Random Walk Based Methods

Random walks can be used for approximating many properties in the graph such as node centrality [26] and similarity [27]. Besides, they are especially available to observe the graph completely. It's also useful when the graph is too large to measure in its entirety. Most random walk based methods adopt a neural language model (Skip-Gram) [28] for graph embedding. DeepWalk and Node2vec are two classic examples using random walks on graphs to obtain node representation.

- **DeepWalk** [29] samples a set of paths from graph using truncated random walk, i.e., uniformly sample the neighbors of the last visited node until the maximum length is reached. Each path sampled from the graph corresponds to a sentence from the corpus, where a node corresponds to a word. Then the Skip-Gram model is applied on the paths to maximize the probability of observing a node's neighborhood conditioned on its embedding. Therefore, nodes with similar neighborhoods are likely to share similar embedding.

- **Node2vec** [30] preserves higher-order proximity between nodes by maximizing the probability of occurrence of subsequent nodes in fixed length random walks. Node2vec employs biased random walks that balance breadth-first (BFS) and depth-first (DFS) sampling. Therefore, it can produce higher quality and more informative embeddings than DeepWalk. And it's important to choose a proper balance between BFS and DFS to preserve community structure based on homophily as well as structural equivalence between nodes.

7.3.3   Deep Learning Based Methods

Deep learning has shown outstanding performance in a wide variety of research fields and there is a deluge of deep neural networks based methods designed for graph embedding graph [31–33]. Graph convolutional networks (GCNs) can tackle the problem of computation expensive for large sparse graphs by defining convolution operators on the graph. In particular, GCN based autoencoder has been widely used, which aims to minimize the reconstruction error of the output and input by its encoder and decoder. The idea of adopting autoencoder for graph embedding is similar to node proximity matrix factorization in terms of neighborhood preservation. In this chapter, we focus on GCN based autoencoder model.

- **GAE** [34] is a non-probabilistic autoencoder using GCN as encoder to obtain the latent representations of nodes, which can be expressed as $Z = GCN(X, A)$. Then a simple inner product decoder is used to reconstruct the original graph, which is able to be described as $\widehat{A} = \sigma(ZZ^T)$. In order to make the constructed adjacency matrices as similar as possible to the original adjacency matrices, $L = E_{q(Z|X,A)}[log_p(A|Z)]$ is defined to demonstrate the similarity.
- **VGAE** [34] is similar to GAE, which utilizes GCN encoder and inner product decoder. The two models rely on GCN to learn the higher-order dependencies between nodes, and their inputs are adjacency matrices. Besides, VAGE defines the lost function as $L = E_{q(Z|X,A)}[log_p(A|Z)] - KL[q(Z|X, A)||p(Z)]$ to demonstrate the similarity between the constructed adjacency matrices and the original adjacency matrix. The results empirically show that using variational autoencoders can improve performance compared to GAE.

7.3.4   Other Methods

- **LINE** [35] preserves both the local and global network structures modeling node co-occurrence probability and conditional probability. It defines two functions for both first-order and second-order proximity and minimizes the combination of the two. The objective function designed for first-order proximity is similar to the function of Graph Factorization (GF) [23]. For each pairwise nodes, LINE defines two joint probability distributions separately using adjacency matrix and embedding. Then LINE minimizes the Kullback–Leibler (KL) divergence of these two distributions to make the adjacency matrix and dot product of embeddings close. Similarly, the second-order proximity also defines probability distributions and objective function.

7.4 The Proposed Method

7.4.1 Basic Definition

Intuitively, as illustrated in Sect. 7.2, the Ethereum transaction records can be modeled as a graph $G = (V, E)$, where V denotes the set of accounts (nodes) and E represents the transaction records (edges) with transaction time. The temporal network can be divided into several snapshots on the basis of time span ϵ. Therefore, in order to grasp the evolution information of the network structure, it's crucial to consider not only the snapshot at the current time but also the nearby snapshots in time. If using single snapshot information of a graph separately, this simple method can not capture the correlation information of the two snapshots next to each other, resulting in temporal loss. Hence, we define TSSN to formulate our solution and further propose TBW to capture the evolution of the whole network. The detailed description of TSSN is shown in Fig. 7.3. Then, we introduce several significant definitions to facilitate the introduction of TBW in Sect. 7.4.2.

Definition 1 (Time-Series Snapshot Network (TSSN)) Given a graph $G = (V, E)$ with timestamps, it can be divided into several independent snapshots $G_1, \cdots, G_T$, where $G_t = (V_t, E_t)$. Let V_t and E_t be a set of nodes and edges of snapshot G_t, respectively, in the timespan $[t\epsilon, (t + 1)\epsilon)$, where ϵ is the time interval, with time order $t \in \{0, 1, 2, \cdots\}$. All snapshots are sorted by time (ascending), and the same nodes in the successive snapshots are connected in order namely self-connections.

Self-connections in TSSN can connect two consecutive snapshots so that random walk can traverse successive snapshots and capture the relevant information between different snapshots, thus obtaining more informative embeddings. In addition, we provide time accessibility for the edge between each pairwise nodes and every two time slices linked nodes. For each edge in TSSN, we define $e = (u, v, w, t): \forall e \in E$, $Src(e) = u$, $Dst(e) = v$, $W(e) = w$, $T(e) = t$, where u is the source node, v is the target node, w is the edge's weight (number of transactions) and t is the edge's time accessibility. Let $\eta_+ : \mathbb{R} \to \mathbb{Z}^+$ be a function that maps each node to an index according to the time order t, e.g., for a given node u in snapshot G_i, we have $\eta_+(u) = i$. Therefore, the time accessibility of each edge (u, v) can be expressed as $T(e) = \eta_+(v) - \eta_+(u)$, where v is the target node and u is the source node, i.e., v is accessible from u if and only if the corresponding $T(e) \geq 0$. Then, we define *successive edges* as follows.

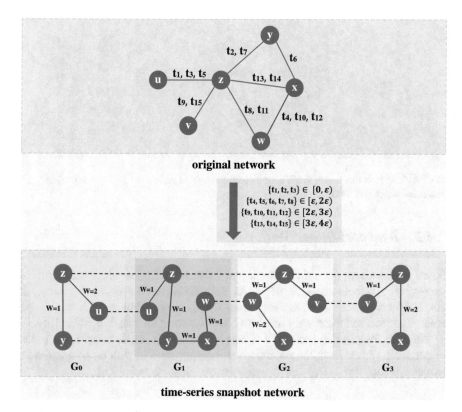

Fig. 7.3 The detailed description of TSSN. The original network is divided into multiple independent snapshots according to time span. And the same nodes in the successive snapshots are connected in TSSN. Dashed lines denote the self-connections, and solid lines denote the connection of node pair in the same snapshot

Definition 2 (Successive Edges) Given a graph $G = (V, E)$, the set of successive edges for a node v is defined as follows:

$$L_t(v) = \{e \mid Src(e) = v, \ T(e) \geq 0\}$$

Figure 7.4 shows an example of successive edge. Consider a random walk that just traversed edge e_{i-1} and is now stopping at node v_i. The next node v_{i+1} of the random walk is decided by selecting a valid successive edge $e_i \in L(v_i)$. The set of successive edges plays the role of a candidate for walkers to select possible successors. Therefore, we propose a joint transfer probability for each successive edge $e \in L_t(v)$, which is composed of the static edge weight, the structural transition probability, and the temporal transition probability. According to the particularity of the dataset, we can expand the transition probability. For instance, in open source software (OSS), the role-based transition probability can be proposed according to the role's real identity.

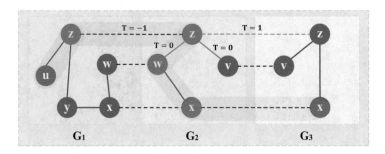

Fig. 7.4 The blue block represents a walk path starting in node u in G_1. The set of node z's accessible edges is denoted as $L_t(z)$

7.4.2 Temporal Biased Walk

Based on the above definitions, we design a second-order neighborhood sampling strategy s to select accessible edges for each node. A unique search strategy is adapted for each node in the network according to its static link weights, node label tendency, structural transition probability, and temporal information probability.

Traditionally, we can perform a simple random walk for an initial node v. Let n_i denote the ith node in the walk sequence $N_s(v)$, starting with $n_0 = v$ and each successive edge $e \in L_t(n_i)$ can be assigned the selection probability:

$$P(e) = \frac{W(e)}{\sum_{e' \in L_t(c)} W(e')} \tag{7.1}$$

where $W(e)$ is the weight value between node c and x, and $L_t(c)$ denotes the valid successive edges of node c. The number of transactions (edge weight) exhibits some differences per account in the Ethereum transaction network. Therefore, we can use this simplest way to bias our temporal random walk, which is to sample the next node based on the static edge weight $W(e)$.

The simplest way to bias our random walks would be to sample the next node according to the static edge weight $W(e)$. However, this simplest random walk fails in explaining the network structure and exploring different types of neighbors in the whole network. Moreover, when the links are relatively sparse in the network, this strategy may get affected easily. The Node2vec [30] method develops a random walk that can explore neighbors in a mixed fashion of BFS and DFS. However, this strategy ignores important time information, which may contain the evolution of the network structure. And it further leads to incomplete network representations.

Intuitively, for a given node v we can divide its first-order neighbors into three categories: inward nodes, outward nodes, and return node. The return node denotes the last selected node. An inward (or outward) node indicates whether there is a link with the return node. If the node has a link with the return node, it is an inward node, otherwise an outward node. Different types of nodes play an important role in the network, thus we need to consider nodes' roles for each node's next walk. Besides, when performing walks, the time accessibility of each node pair is significant which can reflect the evolution of the network structure. Therefore, we propose a second-order random walk method as follows.

Structural Transition Probability   We define the structural transition probability with return parameter r and in-out parameter q similar to Node2vec [30]. For successive edges $e \in L_t(c)$, we set the unnormalized structural transition probability to $P_S(e) = \alpha(e) \cdot W(e)$ with

$$\alpha(e) = P(n_{i+1} = x \mid n_i = c) = \begin{cases} 1/q, & d_{tx} = 2 \\ 1, & d_{tx} = 1 \\ 1/r, & d_{tx} = 0 \end{cases} \tag{7.2}$$

where d_{tx} denotes the shortest path distance between last selected nodes t and next selected node x, and d_{tx} must be one of $\{0,1,2\}$. It is worth noticing that the initial in-out parameter q and return parameter r jointly determine every node's search direction. Our method, like [30, 36], uses the return parameter r and in-out parameter q to control how fast the walk explores and leaves the neighborhood of starting node v. Therefore, our method can explore more different types of nodes to improve representation ability.

Temporal Transition Probability   Besides structural features, the temporal information also counts for a lot in the node representation learning. When we divide the whole network into different time slices according to the time span, each time slice represents a part of the network structure and the gradual change of time slice reflects the evolution process of the network. Ignoring the correlation information that exists between two snapshots at consecutive time steps may cause the loss of temporal information. Hence, we propose temporal transition probability to capture node's behavior changes in different snapshots. In this case, the probability of selecting each edge $e \in L_t(c)$ can be given as:

$$P_T(e) = \frac{\varphi(e)}{\sum_{e' \in L_t(c)} \varphi(e')} \tag{7.3}$$

where $\varphi(e)$ is expressed as

$$\varphi(e) = \begin{cases} \alpha, & T(e) > 0 \\ 1 - \alpha, & T(e) = 0 \end{cases} \tag{7.4}$$

Here, the temporal bias $\alpha(0.1 \leq \alpha \leq 0.9)$ decides whether the temporal walk resides on the current snapshot or transfer to the next.

Joint Transition Probability   Furthermore, we normalize the aforementioned structural transition probability and temporal transition probability and then combine them as one. We set the unnormalized transition probability to $P(e)$, and then normalize it to the final transition probability for each edge $e \in L_t(c)$, where

$$P(e) = P_S(e)\, P_T(e). \tag{7.5}$$

We propose a second-order neighborhood sampling strategy s which can help each node find a suitable search direction and get its optimal temporal successive edges. In joint transition probability, in-out parameter q, return parameter r, and temporal bias α jointly determine the search direction.

The return parameter r mainly controls the probability of the source node revisiting the return node. When r is low, it would keep the walk close to the source node. On the other hand, setting it to a high value ensures that the walk is less likely to the already visited node. The parameter q prefers to consider searching for different types of inward and outward nodes. When $q > 1$, the next walk of the source node is more inclined to return to the source node, which is more like a local exploration like the BFS behavior. When $q < 1$, the source node is more likely to walk away from the source node. This method can make the source node explore a wider range of nodes, which is a kind of approximate DFS behavior. By adjusting the parameter q, we allow our search process to combine BFS with DFS. On the whole, the in-out parameter q and return parameter r control the search direction in spatial domain simultaneously. Temporal bias α decides the temporal search orientation: reside on current snapshot or move to the next snapshot. If α is small, the temporal walk is more inclined to stay in the current snapshot, otherwise favors edges appearing in the future snapshot. This is helpful to explore the changes of node interaction in very different time periods with the development of networks. These samplings help to search for various nodes both in the spatial and temporal domains.

7.4.3 Learning Temporal Graph Embeddings

Our goal is to obtain a mapping function $\Phi : V \to \mathbb{R}^d$, which maps a given node to a d-dimensional representation. For a node $v \in V$, let $N_s(v)$ denote the set of temporal neighbors that are generated according to the search strategy s, and $\Phi_t(v)$ is the representation of node v in snapshot G_t. Our objective function is to obtain a d-dimensional representation for a given node v and the function maximizes the log-probability of observing $N_s(v)$ and historical embedding $\Phi_t(v)$ for the node v conditioned on its representation:

$$\max_{\Phi} \sum_{v \in V} \log(Pr(N_s(v), \Phi_t(v) \mid \Phi(v))). \tag{7.6}$$

We assume that the temporal neighbors in $N_s(v)$ and the nodes' historical representations $\Phi_t(v)$ are independent of each other. Accordingly, we factorize the formula:

$$\log(Pr(N_s(v), \Phi_t(v) \mid \Phi(v)))$$
$$= \log(\prod_{u_i \in N_s(v)} Pr(u_i \mid \Phi(v))) + \log(Pr(\Phi_t(v) \mid \Phi(v))). \qquad (7.7)$$

Based on the network analysis, we can see that the likelihood of observing a source node is independent of observing any other and the definition of neighborhood nodes is symmetric [30]. Therefore, we factorize the likelihood of observing temporal neighbors and model the likelihood of every source-neighborhood node pair as a softmax unit that is parametrized by a dot product of their mapping features. Learning representations using random walk has proved to measure better graph proximity, and thereby improving the performance [21, 37]. Hence, we use random walk to learn the conditional probability of observing a node u_i given the learned representation $\Phi(v)$ as follows:

$$Pr(u_i \mid \Phi(v)) = \frac{\exp(\Phi(u_i) \, \Phi(v))}{\sum_{n \in V} \exp(\Phi(n) \, \Phi(v))}, \qquad (7.8)$$

where $u_i \in N_s(v)$ is the ith neighbor of node v. With the above hypothesis, the objective function in Eq. (7.6) is simplified to:

$$\max_f \sum_{v \in V} \log(\prod_{u_i \in N_s(v)} \frac{\exp(\Phi(u_i) \, \Phi(v))}{\sum_{n \in V} \exp(\Phi(n) \, \Phi(v))})$$
$$+ \log(Pr(f_t(v) \mid f(v))). \qquad (7.9)$$

While the above seems to just consider the process of network topological properties, it actually takes into account the structure of network at different times and thus reflects the evolution process of network topology properties. Due to the nonlinear nature of real-world networks, we define a novel search strategy s that samples different temporal neighbors of a given source node v. The temporal neighbors $N_s(v)$ are not restricted to just the nearest neighbors but also have vastly structural similarity with the source node in spatial and temporal domains simultaneously. Considering the complexity of the objective function, we use negative sampling strategy to approximate it [38]. The stochastic gradient descent (SGD) [39] method is used to iteratively update the objective function.

The pseudocode for time-preserving embeddings in TSSN is given in Algorithms 1. We summarize the second-order random walk strategy (TBW) in Algorithm 2. Our procedure in Algorithm 1 generalizes the Skip-Gram architecture to learn embeddings in TSSN. The three phases of TBW, i.e., preprocessing to compute joint transition probability, random walk simulations, and optimization using SGD, are executed sequentially. Each phase is parallelizable and can be executed asynchronously, which contributes to the overall scalability of TBW. In addition, since the temporal walks can be used as the input vector of the neural network, TBW can be easily used in deep graph models. There are many random walk methods in TSSN that can be adapted because it does not depend on any specific method.

Algorithm 1: Time-preserving embedding framework

Input: temrpoal graph $G = (V, E)$, return r, in-out q, temporal bias α, time span ϵ, dimension d, walks per node w, walk length l, window size k
Output: $f(v)$ for $\forall v \in V$

1 Initialize set of temporal walks N_s to $\emptyset$
2 $G' = \text{CreateTSSN}(G, \epsilon)$
3 **for** *iter=1 to w* **do**
4   **for** *all node* $v \in V$ **do**
5          $P = \text{PrecomputeTransitionProbability}(G', r, q, \alpha)$
6          $walk = \text{TemporalBiasedWalk}(G', v, l, P)$
7   Append $walk$ to N_s
8   **end**
9 **end**
10 $f = \text{StochasticGradientDescent}(k, d, N_s)$
11 return $\Phi \in \mathbb{R}^{|V| \times d}$

Algorithm 2: Temporal biased walk

Input: time-series snapshot network G', start node u, walk length l, transition probability P
Output: temporal walk $walk$

1 Initialize $walk$ to $[u]$
2 **for** *iter=1 to l* **do**
3      $curr = walk[-1]$
4      $e = \text{AliasNodeSample}(curr, P)$
5   Append $Dst(e)$ to $walk$
6 **end**
7 return $walk$

7.5 Experiment

7.5.1 Node Classification

Phishing scam is a new type of cybercrime with the emergence of blockchain technology. It is reported to account for more than 50% of all cyber-crimes in Ethereum since 2017 [40]. Therefore, it's important to find out phishing accounts to maintain the security of online business. In this subsection, we conduct node classification experiment on Ethereum to classify labeled phishing nodes and unlabeled nodes (treated as non-phishing nodes). We compare the effectiveness of embedding methods by using the generated embeddings as node features to classify the nodes. The node features are input to a one-vs-rest logistic regression using the LIBLINEAR library. Experiments are repeated 5 times and the mean with confidence interval is reported.

7.5.1.1 Evaluation Metrics

- **F1-Score** is a measure of the test's accuracy and the harmonic mean of precision and recall. The highest possible value of F1 is 1, indicating perfect precision and recall, and the lowest possible value is 0 if either the precision or the recall is zero. It is defined as:

$$F1 = \frac{2PR}{P + R}$$

where P and R are the precision and recall, respectively. The precision is the number of correctly identified positive results divided by the number of all positive results, including those not identified correctly, and the recall is the number of correctly identified positive results divided by the number of all samples that should have been identified as positive.

7.5.1.2 Experimental Results

We randomly sample 10% to 90% of nodes as training data and evaluate the performance on the remaining nodes. Figure 7.5a shows the performance, from which one can see that our method TBW significantly outperforms baselines varying different test ratio. And TBW and Node2vec have similar performance curves since both of them preserve homophily as well as structural equivalence between nodes, suggesting their effectiveness in node classification. We further investigate the impact of embedding dimensions on node classification, as shown in Fig. 7.5b, from which one can observe that classification performance often saturates or deteriorates with the increase of embedding dimension. This may be because with a higher dimension these embedding methods overfit on the labeled nodes and are unable to predict the label of remaining nodes. With a couple of exceptions, as the number of dimension increases, the F1-Score value increases. This is intuitive as higher number of dimension is capable of storing more information. We also observe that LINE achieves the best performance with 16 dimension. This may be because the embedding model overfits in higher-order proximity.

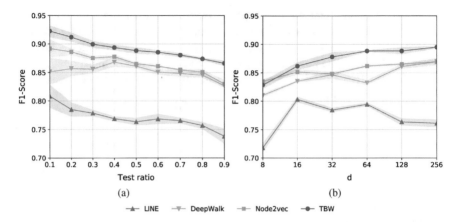

Fig. 7.5 Performance of node classification. (**a**) F1-Score of node classification varying the train-test split ratio (dimension of embedding is 128). (**b**) F1-Score of node classification varying the number of dimension (the train-test split is 50%)

7.5.2   Link Prediction

The task of link prediction aims to predict the occurrence of links in a given graph based on observed information. In this subsection, we treat the predictive account transactions as a link prediction task in Ethereum. Before the experiments, we hide a certain fraction of accounts' connections in the network, and our goal is to trace these missing connections via link prediction. We first randomly hide 20% of links in the original network as the ground truth and use the remaining to train all graph embedding models. The test set consists of two parts, one of which is all the hidden edges, and the other is the unconnected pairwise nodes sampled randomly as the negative samples. After learning embedding for each node, we use hadamard[1] operation on the learned embedding vectors of pairwise nodes to compute the feature vector for the corresponding edge. For all embedding techniques, we implement experiments by using a one-vs-rest logistic regression classifier. Experiments are repeated for 5 random seed initializations and the average performance is reported.

7.5.2.1   Evaluation Metrics

- **AUC** can be interpreted as the probability that a randomly chosen missing link is given a higher score than a randomly chosen nonexistent link. If among n independent comparisons, there are n' times that the missing link gets a higher

---

[1]$[f(u) \cdot f(v)]_i = f_i(u) * f_i(v)$.

score and n'' times they get the same score, the AUC value is

$$\text{AUC} = \frac{n' + 0.5n''}{n}$$

If all the scores are generated from an independent and identical distribution, the AUC value should be about 0.5. Therefore, the degree to which the value exceeds 0.5 indicates how much better the algorithm performs than pure chance.

- **Precision** is defined as the ratio of relevant items selected to the number of items selected. That is to say, if we take the top-L links as the predicted ones, among which L_r links are right, then the Precision equals L_r/L. Clearly, higher precision means higher prediction accuracy.

7.5.2.2 Experimental Results

The experimental results of link prediction with 128-dimensional embeddings are given in Table 7.2. Besides, Fig. 7.6 shows the average precision and AUC results for link prediction for each dimension, from which one can see that the performance of embedding methods highly depends on the dataset and embedding dimension. Specifically, HOPE achieves good performance when dimension is small but performs poorly when dimension increases. A reasonable explanation is that the model overfits on the observed links and fails in predicting unobserved links. GF and LINE achieve poor performance on most networks, indicating that preserving higher-order proximity is not conducive to predicting unobserved links. Here, the random walk based method outperforms other methods, indicating that random walks are especially useful when approximate node centrality and similarity in Ethereum transaction networks.

Typically, similarity indices are used for link prediction to estimate the likelihood of a link being present. Basic similarity indices include: *Common Neighbors*, *Jaccard Index*, *Adamic-Adar Index*, *Resource Allocation Index*. These indicators are detailed in Appendix 7.7. Similarly, GCN based methods have shown outstanding performance in link prediction. However, comparing with similarity-based link pre-

Table 7.2 Performance for link prediction with 128 dimensional embeddings. The best results are marked in bold

Metrics	EthereumG1		EthereumG2		EthereumG3	
	AP	AUC	AP	AUC	AP	AUC
GF	0.7827	0.6821	0.7377	0.7050	0.7946	0.6662
HOPE	0.7698	0.6578	0.8300	0.7580	**0.8714**	0.8089
LINE	0.7761	0.7521	0.8627	0.8237	0.6371	0.6370
DeepWalk	0.6159	0.6637	0.6138	0.6307	0.7755	0.8024
Node2vec	0.6877	0.7149	0.6939	0.6990	0.8239	0.8501
TBW	**0.8553**	**0.8700**	**0.8898**	**0.8818**	0.8622	**0.8687**

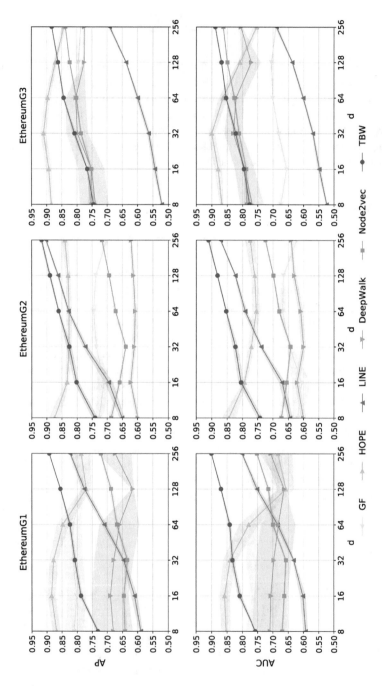

Fig. 7.6 Performance of link prediction for different datasets with varying dimensions

Table 7.3 The results of comparison with similarity-based and GCN based methods for link prediction. The best results are marked in bold

Metrics	EthereumG1		EthereumG2		EthereumG3	
	AP	AUC	AP	AUC	AP	AUC
CN	0.6907	0.6848	0.5134	0.4921	0.6826	0.6881
Jaccard	0.5088	0.6097	0.4523	0.4588	0.6234	0.6800
AA	0.7367	0.7002	0.5909	0.5099	0.6979	0.6912
RA	0.7378	0.7007	0.5909	0.5099	0.6986	0.6914
GAE	0.7911	0.6752	0.5828	0.3729	0.8703	0.7885
VGAE	0.8179	0.7184	0.6683	0.4719	**0.8934**	0.8278
TBW	**0.8553**	**0.8700**	**0.8898**	**0.8818**	0.8622	**0.8687**

diction methods and GCN based methods, the effectiveness of our proposed method is yet to be verified. To study the competitiveness of our proposed method, we conduct experiments of various similarity-based link prediction methods and GCN based methods with 128-dimensional embeddings. The average precision and AUC results are presented in Table 7.3. We can observe that TBW achieves consistently and significantly better performance over similarity-based link prediction methods, which is reasonable since our method with flexible walking strategies are able to learn the similarity between nodes more effectively.

7.6 Conclusion

In this chapter, we construct TSSN to retain both temporal and structural information of Ethereum transaction network as much as possible and present temporal biased walk (TBW) to make phishing detection and transaction tracking by leveraging embeddings learned from structural properties and temporal information. Particularly, we transform phishing detection and transaction tracking into node classification and link prediction from a network perspective. Furthermore, we implement the proposed embedding method on realistic Ethereum transaction network for node classification and link prediction with practical relevance. We compare our method with a number of graph embedding techniques. The experimental results demonstrate the effectiveness of the proposed TBW and indicate that TSSN can more comprehensively represent the temporal and structural properties of the Ethereum transaction network.

Though Ethereum transaction records are publicly available, it's still relatively unexplored till now. And the effects of our proposed method on other realistic downstream tasks remain to be verified. For future work, we plan to apply deep learning methods to expand our methods or extend the current framework to analyze more illegal activities on Ethereum and create a safe trading environment for Ethereum.

7.7 Appendix

7.7.1 Similarity Indices

- **Common Neighbours (CN)**
  It is defined as [6]:

  $$s_{ij}^{CN} = |\Gamma(i) - \Gamma(j)|$$

  where $\Gamma(i)$ denotes the set of neighbors of i and $|x|$ is the cardinality of the set x. In common sense, two nodes i and j are more likely to have a link if they have many common neighbors. It is obvious that $s_{ij} = (A^2)_{ij}$, where A is the adjacency matrix: $A_{ij} = 1$ if i and j are directly connected otherwise $A_{ij} = 0$.
- **Jaccard Index (Jaccard)**
  It is defined as [41]:

  $$s_{ij}^{Jaccard} = \frac{|\Gamma(i) \bigcap \Gamma(j)|}{|\Gamma(i) \bigcup \Gamma(j)|}$$

  Jaccard is a classical statistical parameter used to compare the similarity or diversity of sample sets.
- **Adamic-Adar Index (AA)**
  It is defined as [42]:

  $$s_{ij}^{AA} = \sum_{z \in \Gamma(i) \bigcap \Gamma(j)} \frac{1}{\log k_z}$$

  This index refines the simple counting of common neighbors by assigning the less-connected neighbors more weight. For example, most individuals may know a famous man, but they themselves may not know each other.
- **Resource Allocation Index (RA)**
  It is defined as [43]:

  $$s_{ij}^{RA} = \sum_{z \in \Gamma(i) \bigcap \Gamma(j)} \frac{1}{k_z}$$

  RA index is close to AA, but punish more on their common neighbors of higher degree and this is motivated by the resource allocation dynamics on complex networks. In some cases, RA performs better than AA in link prediction.

References

1. Swan, M.: Blockchain: Blueprint for a New Economy. O'Reilly Media, Inc., Sebastopol (2015)
2. Holub, A., O'Connor, J.: Coinhoarder: tracking a Ukrainian bitcoin phishing ring DNS style. In: 2018 APWG Symposium on Electronic Crime Research (eCrime), pp. 1–5. IEEE, Piscataway (2018)
3. Russon, M.A.: Ethereum under siege: scammers make $700000 in 6 days from slack and Reddit phishing attacks (2017)
4. Khonji, M., Iraqi, Y., Jones, A.: Phishing detection: a literature survey. IEEE Commun. Surv. Tut. 15(4), 2091–2121 (2013)
5. Rossi, R., Neville, J.: Time-evolving relational classification and ensemble methods. In: Pacific-Asia Conference on Knowledge Discovery and Data Mining, pp. 1–13. Springer, Berlin (2012)
6. Lü, L., Zhou, T.: Link prediction in complex networks: a survey. Phys. A Stat. Mech. Appl. 390(6), 1150–1170 (2011)
7. Fortunato, S.: Community detection in graphs. Phys. Rep. 486(3–5), 75–174 (2010)
8. Ron, D., Shamir, A.: Quantitative analysis of the full bitcoin transaction graph. In: International Conference on Financial Cryptography and Data Security, pp. 6–24. Springer, Berlin (2013)
9. Jiang, Z., Liang, J.: Cryptocurrency portfolio management with deep reinforcement learning. In: 2017 Intelligent Systems Conference (IntelliSys), pp. 905–913. IEEE, Piscataway (2017)
10. Chen, W., Zheng, Z., Cui, J., Ngai, E., Zheng, P., Zhou, Y.: Detecting Ponzi schemes on ethereum: Towards healthier blockchain technology. In: Proceedings of the 2018 World Wide Web Conference, pp. 1409–1418 (2018)
11. Liang, J., Li, L., Zeng, D.: Evolutionary dynamics of cryptocurrency transaction networks: an empirical study. PLoS One 13(8), e0202202 (2018)
12. Wu, J., Liu, J., Chen, W., Huang, H., Zheng, Z., Zhang, Y.: Detecting mixing services via mining bitcoin transaction network with hybrid motifs. Preprint. arXiv:2001.05233 (2020)
13. Wu, J., Yuan, Q., Lin, D., You, W., Chen, W., Chen, C., Zheng, Z.: Who are the phishers? Phishing scam detection on ethereum via network embedding. IEEE Trans. Syst. Man Cyber. Syst. 1–11 (2020)
14. Yuan, Q., Huang, B., Zhang, J., Wu, J., H. Zhang, Zhang, X.: Detecting phishing scams on ethereum based on transaction records. In: 2020 IEEE International Symposium on Circuits and Systems (ISCAS), pp. 1–5. IEEE, Piscataway (2020)
15. Wu, J., Lin, D., Zheng, Z., Yuan, Q.: T-edge: temporal weighted multidigraph embedding for ethereum transaction network analysis. Preprint. arXiv:1905.08038 (2019)
16. Wood, G., et al.: Ethereum: a secure decentralised generalised transaction ledger. Ethereum Project Yellow Paper 151(2014), 1–32 (2014)
17. Weili, C., Zibin, Z.: Blockchain data analysis: a review of status, trends and challenges. J. Comput. Res. Develop. 55(9), 1853 (2018)
18. Wang, S., Ouyang, L., Yuan, Y., Ni, X., Han, X., Wang, F.-Y.: Blockchain-enabled smart contracts: architecture, applications, and future trends. IEEE Trans. Syst. Man Cyber. Syst. 49(11), 2266–2277 (2019)
19. Chen, T., Zhu, Y., Li, Z., Chen, J., Li, X., Luo, X., Lin, X., Zhange, X.: Understanding ethereum via graph analysis. In: IEEE INFOCOM 2018-IEEE Conference on Computer Communications, pp. 1484–1492. IEEE, Piscataway (2018)
20. Lin, D., Wu, J., Yuan, Q., Zheng, Z.: Modeling and understanding ethereum transaction records via a complex network approach. IEEE Trans. Circuits Syst. Express Briefs 67(11), 2737–2741 (2020)
21. Goyal, P., Ferrara, E.: Graph embedding techniques, applications, and performance: a survey. Knowl. Based Syst. 151, 78–94 (2018)
22. Cai, H., Zheng, V.W., Chang, K.C.-C.: A comprehensive survey of graph embedding: Problems, techniques, and applications. IEEE Trans. Knowl. Data Eng. 30(9), 1616–1637 (2018)

23. Ahmed, A., Shervashidze, N., Narayanamurthy, S., Josifovski, V., Smola, A.J.: Distributed large-scale natural graph factorization. In: Proceedings of the 22nd International Conference on World Wide Web, pp. 37–48 (2013)
24. Ou, M., Cui, P., Pei, J., Zhang, Z., Zhu, W.: Asymmetric transitivity preserving graph embedding. In: Proceedings of the 22nd ACM SIGKDD International Conference on Knowledge Discovery and Data Mining, pp. 1105–1114 (2016)
25. Van Loan, C.F.: Generalizing the singular value decomposition. SIAM J. Numer. Anal. **13**(1), 76–83 (1976)
26. Newman, M.E.J.: A measure of betweenness centrality based on random walks. Soc. Net. **27**(1), 39–54 (2005)
27. Fouss, F., Pirotte, A., Renders, J.-M., Saerens, M.: Random-walk computation of similarities between nodes of a graph with application to collaborative recommendation. IEEE Trans. Knowl. Data Eng. **19**(3), 355–369 (2007)
28. Mikolov, T., Chen, K., Corrado, G., Dean, J.: Efficient estimation of word representations in vector space. Preprint. arXiv:1301.3781 (2013)
29. Perozzi, B., Al-Rfou, R., Skiena, S.: Deepwalk: online learning of social representations. In: Proceedings of the 20th ACM SIGKDD International Conference on Knowledge Discovery and Data Mining, pp. 701–710 (2014)
30. Grover, A., Leskovec, J.: node2vec: scalable feature learning for networks. In: Proceedings of the 22nd ACM SIGKDD International Conference on Knowledge Discovery and Data Mining, pp. 855–864 (2016)
31. Wang, D., Cui, P., Zhu, W.: Structural deep network embedding. In: Proceedings of the 22nd ACM SIGKDD International Conference on Knowledge Discovery and Data Mining, pp. 1225–1234 (2016)
32. Cao, S., Lu, W., Xu, Q.: Deep neural networks for learning graph representations. In: Proceedings of the AAAI Conference on Artificial Intelligence, vol. 16, pp. 1145–1152 (2016)
33. Niepert, M., Ahmed, M., Kutzkov, K.: Learning convolutional neural networks for graphs. In: International Conference on Machine Learning, pp. 2014–2023 (2016)
34. Kipf, T.N., Welling, M.: Variational graph auto-encoders. Preprint. arXiv:1611.07308 (2016)
35. Tang, J., Qu, M., Wang, M., Zhang, M., Yan, J., Mei, Q.: Line: large-scale information network embedding. In: Proceedings of the 24th International Conference on World Wide Web, pp. 1067–1077 (2015)
36. Chen, J., Wu, Y., Xu, X., Zheng, H., Ruan, Z., Xuan, Q.: Pso-ane: adaptive network embedding with particle swarm optimization. IEEE Trans. Comput. Soc. Syst. **6**(4), 649–659 (2019)
37. Hamilton, W.L., Ying, R., Leskovec, J.: Representation learning on graphs: methods and applications. Preprint. arXiv:1709.05584 (2017)
38. Mikolov, T., Sutskever, I., Chen, K., Corrado, G.S., Dean, J.: Distributed representations of words and phrases and their compositionality. In: Advances in Neural Information Processing Systems, pp. 3111–3119 (2013)
39. Bottou, L.: Large-scale machine learning with stochastic gradient descent. In: Proceedings of COMPSTAT'2010, pp. 177–186. Springer, Berlin (2010)
40. Conti, M., Kumar, E.S., Lal, C., Ruj, S.: A survey on security and privacy issues of bitcoin. IEEE Commun. Surv. Tutorials **20**(4), 3416–3452 (2018)
41. Jaccard, P.: Étude comparative de la distribution florale dans une portion des alpes et des jura. Bull. Soc. Vaudoise Sci. Nat. **37**, 547–579 (1901)
42. Adamic, L.A., Adar, E.: Friends and neighbors on the web. Soc. Netw. **25**(3), 211–230 (2003)
43. Zhou, T., L. Lü, Zhang, Y.-C.: Predicting missing links via local information. Euro. Phys. J. B **71**(4), 623–630 (2009)

Chapter 8
Find Your Meal Pal: A Case Study on Yelp Network

Jian Zhang, Jie Xia, Laijian Li, Binda Shen, Jinhuan Wang, and Qi Xuan

Abstract Yelp is an online website for reviewing restaurants, stores and so on. Users can grade restaurants and share their dining experiences through text and photos. Along with the social relationships between users, Yelp enables a recommendation engine to make precise restaurants recommendations, which improves the user experience of the website and promote the revenues of restaurants. In this chapter, we focus on the Yelp friend network to make friends recommendation through random forest (RF) and variational graph auto-encoder (VGAE). The former method assembles multiple handcraft node similarity indices while the latter one could automatically learn network structural features. Moreover, we construct a co-foraging network to analyze the co-foraging patterns on Yelp and recommend potential meal pals to users. The experiments show the effectiveness of the recommendation methods and reveal the possibility of applying link prediction approaches to Yelp data analysis.

8.1 Introduction

Founded in 2004, Yelp[1] is a website where users grade restaurants and share dining experiences. Not only limited to restaurants' information, Yelp also offers the information of shopping centers, hotels and tourist attractions, covering the major aspects of daily life. More than 178 millions visitors per month across different

---

[1] https://www.yelp.com.

J. Zhang · J. Wang · Q. Xuan (✉)
Institute of Cyberspace Security, Zhejiang University of Technology, Hangzhou, China

College of Information Engineering, Zhejiang University of Technology, Hangzhou, China
e-mail: xuanqi@zjut.edu.cn

J. Xia · L. Li · B. Shen
College of Information Engineering, Zhejiang University of Technology, Hangzhou, China

© The Author(s), under exclusive license to Springer Nature Singapore Pte Ltd. 2021
Q. Xuan et al. (eds.), *Graph Data Mining*, Big Data Management,
https://doi.org/10.1007/978-981-16-2609-8_8

platforms and 184 million reviews worldwide make Yelp the 44th most visited website in the US based on the data collected by Alexa. Customers give ratings and detailed reviews to businesses according to their own offline experiences, helping other users to make decisions based on their preferences. And businesses could present their features like special flavors on the website to attract more customers. Connecting users and local businesses, many other websites, such as *Foursquare*, *UrbanSpoon* and *Dianping*, also offer similar services.

The website, Yelp.com, is enriched with data of the basic information of businesses as well as users, including reviews of businesses and interactions between users. Motivated by such numerous data, Yelp enables a recommendation engine to aid users to find their preferred services and promote the sales of registered merchants. Moreover, Yelp itself has held 12 rounds of competitions, namely *Yelp Dataset Challenge*, to encourage researchers to make academic contributions. A bunch of researches have been carried out on Yelp dataset and many of them focus on the reviews. Yang et al. [1] find the length and readability of a textual review affect its usefulness and the aspects of a graphical review have a positive influence on its enjoyment. And Huang et al. [2] propose to improve restaurants according the subtopics of their reviews.

Apart from diverse text and image data, the interactions between users as well as the interactions between users and restaurants are also valuable. By modeling the interactions as different networks, network analysis techniques could be applied in various applications. Cervellini et al. [3] propose to find trendsetters on Yelp by ranking the nodes in the Yelp social network. Trendsetters are those who reviewed restaurants before they reach their peak popularity [4]. They have great influence on their friends and thus making the identification of trendsetters contribute to the popularity of restaurants. Not limited to friendship networks, there are various kinds of networks modeling different kinds of relations. Using the geographical data of restaurant reviews, Xuan et al. [5] construct a geographical foraging network and a taste foraging network which could characterize a patron's foraging behaviors. And Fu et al. [6] model the friendship and co-foraging behaviors of users as a two-layer network and infer the possibility whether two users will have meals together. If restaurants, reviews, ratings and locations are also considered as nodes, the complex relationships between different kinds of entities could be modeled as a heterogeneous network. Meta-Path [7] is then proposed to make personalized recommendation.

In fact, making recommendations on Yelp can be considered as predicting links between two entities. In this chapter, we focus on two networks, a friendship network and a co-foraging network, to make friends and meal pals recommendations in the way of link prediction. As we have briefly reviewed in Chap. 2, link prediction could be achieved by similarity index-based methods [8] and random walk-based approaches [9]. And newly emerged graph neural networks [10–12] are also applicable to solve the problem. In the experiments, we assembles 11 similarity indices by random forest (RF) to predict links in the above two networks. Also, we compare its performance with different link prediction methods, including node2vec, variational graph auto-encoder (VGAE) and hyper-substructure enhanced link predictor (HELP) introduced in Chap. 2. And due to the interpretability of RF, we then further investigate into the importance of each similarity index to compare the effectiveness of different indices. The results presented in Sect. 8.4 prove the practicability of link prediction in making recommendation on Yelp dataset.

In the following of the chapter, we first give a detailed description of Yelp dataset and the construction of networks in Sect. 8.2. In Sect. 8.3, we introduce the link prediction methods used in this chapter. And the experiments of friends and meal pals recommendation are presented in Sect. 8.4. We conclude the chapter with future works in Sect. 8.6.

8.2   Data Description and Preprocessing

On Yelp, each restaurant is advertised with its name, location, contact information and opening hours. And each one comes with several tags to show their features and an average score denoted by stars which reflects the quality of food and service. The average score is based on all the reviews given by customers. The merchants may also highlight some useful information on their pages. In Fig. 8.1, a restaurant named *Milk Jar Cookies* is tagged with *Desserts*, *Bakeries* and *Coffee & Tea*. And it has 4.5 stars and 1,575 reviews in total.

As for the reviews, they are posted by customers right on the pages of restaurants. Figure 8.2 gives an example review of *Milk Jar Cookies*. The review contains a detailed description of the corresponding consumer's experience. The text detailedly describes the reviewer's feeling about the food and the services. Also, a rating denoted by stars is given to the restaurant. Along with the text, the reviewer may also attach several pictures of the surroundings and the food of the restaurant. Other users interested in the restaurant could have a better knowledge about the restaurant based on the review, not just based on the information provided by the restaurant. And they would tag the review with **Useful**, **Funny** or **Cool** according to the content of the review. Apart from the information of restaurants and the reviews, the profiles of users and their social relationships are also provided on Yelp. And the friendships between users are what we focus on here.

Fig. 8.1 An overview of a restaurant on Yelp

Mia K. `Elite '2021`
Los Angeles, CA
215   621   1254

★★★★★ 1/27/2021

2 photos

Got a box of these deeelicious cookies as a gift. The packaging was festively rustic. The cookies are huge, soft, and perfectly sweet. My favorites were Chocolate Mint, seasonal Butterscotch, Chocolate Peanut Butter, and S'mores.

Even though the cookies are said to be fresh for 3-4 days, I found them to still taste great for longer.

Inside is so quaint, reminds me of a relaxing throwback to an ice cream parlor. Must go here for s'more sweet goodness on cheat day!

Useful 2   Funny   Cool 1

Fig. 8.2 An example review on Yelp

We extract the data published by Yelp Dataset Challenge[2] which contains part of the businesses, reviews, and user data on Yelp. In more than six million friendship records, we filter out the users with fewer than 50 history reviews and 500 friends to avoid the affection of inactive users. Then we sample part of the users by breadth-first search and construct an undirected and unweighted network as an example network in this chapter. The final friend network, denoted as G_f, consists of 4030 users connected by 37,493 edges which is visualized in Fig. 8.3.

Based on the friendship network, we further construct a foraging network whose edges represent the co-foraging behavior between the users in G_f. We collect all the 31,959 restaurants that the 4,030 users have ever reviewed. With the assumption that the users who are geographically close are more likely to have meals together, we adopt the density-based spatial clustering of applications with noise (DBSCAN) to divide those restaurants into 11 clusters according to their longitudes and latitudes. Within each cluster, we further divide the restaurants by K-means to ensure that the restaurants within a cluster are geographically close to each other. Finally we get 71 clusters of restaurants. Given two users u and v, they are connected if they are friends in G_f and have ever reviewed at least 10 restaurants in one cluster. It is more reasonable to recommend meal pals within a patron's friends. In this way, we obtain

---

[2]https://www.yelp.com/dataset.

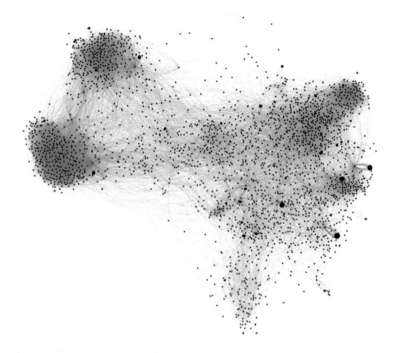

Fig. 8.3 The friendship network of 4030 users on Yelp

the Yelp co-foraging network, denoted by G_c. Apparently, G_c is a subgraph of G_f but the edges have different meanings. G_c consists of 4,030 nodes and 13,014 edges in total.

8.3   Link Prediction Methods

In this section, we briefly introduce the methods we adopt on the Yelp dataset. Apart from the similarity index-based methods and node2vec we have reviewed in previous chapters, we propose to use assemble a plenty of similarity indices to perform link prediction. Also, we use VGAE and HELP to present the application of link prediction on Yelp dataset.

8.3.1   Similarity Indices Assembly

As we have introduced in previous chapters, local/global similarity indices could be used to infer potential links. For example, we could justify whether there is a link between two users according to the common friends they have. In this chapter, we

Table 8.1 The definition of similarity indices

Similarity index	Definition
Common Neighbors (CN)	$s_{ij}^{CN} = \|\Gamma(i) \cap \Gamma(j)\|$
Salton Index (SA)	$s_{ij}^{SA} = \frac{\|\Gamma(i) \cap \Gamma(j)\|}{\sqrt{k_i \times k_j}}$
Jaccard Index (JAC)	$s_{ij}^{JAC} = \frac{\|\Gamma(i) \cap \Gamma(j)\|}{\|\Gamma(i) \cup \Gamma(j)\|}$
Hub Promoted Index (HPI)	$s_{ij}^{HPI} = \frac{\|\Gamma(i) \cap \Gamma(j)\|}{\sqrt{k_i \times k_j}}$
Hub Depressed Index (HDI)	$s_{ij}^{HDI} = \frac{\|\Gamma(i) \cap \Gamma(j)\|}{max(k_i \times k_j)}$
SΦrensen Index (SI)	$s_{ij}^{SI} = \frac{\|\Gamma(i) \cap \Gamma(j)\|}{k_i + k_j}$
Leicht-Holme-Newman Index (LHN)	$s_{ij}^{LHN} = \frac{\|\Gamma(i) \cap \Gamma(j)\|}{k_i \times k_j}$
Adamic-Adar Index (AA)	$s_{ij}^{AA} = \sum_{z \in \Gamma(i) \cap \Gamma(j)} \frac{1}{log(k_z)}$
Resource Allocation Index (RA)	$s_{ij}^{AA} = \sum_{z \in \Gamma(i) \cap \Gamma(j)} \frac{1}{k_z}$
Preferential Attachment Index (PA)	$s_{ij}^{PA} = k_i \times k_j$
Friends-Measure (FM)	$s_{ij}^{FM} = \sum_{u \in \Gamma i} \sum_{v \in \Gamma(j)} \delta(u, v)$
Local Path Index (LP)	$s_{ij}^{LP} = (A^2)_{ij} + \epsilon(A^3)_{ij}$

assembles several similarity indices and adopt random forest to infer the linkage status. In detail, the similarity indices we use are listed in Table 8.1. Given a network, we first calculate the 11 similarity indices for each node pair and then concatenate them together into one vector as the features of corresponding node pair. Afterwards, we adopt the RF model which assembles the 11 similarity indices to perform link prediction.

8.3.2 Variational Graph Auto-Encoder

Variational Graph Auto-Encoder [13] automatically learns network structures without pre-defined heuristics. Different from auto-encoder, it regularizes the distribution of the hidden representations and then generates data from the estimated distribution instead of reconstructing directly from the hidden features. With the assumption that the hidden representations of the network subject to Gauss distribution, VGAE first encodes the input network into lower dimensional representations by applying GCN and then learns the distribution, μ and σ, of the network. μ and σ are the mean and standard variance matrices of the hidden representations, respectively. Mathematically, the process could be modeled as

$$h = GCN(X, A),$$

$$\mu = GCN_{\mu}(h, A),$$

$$log\sigma = GCN_{\sigma}(h, A),$$

(8.1)

where $A \in \mathbb{R}^{N \times N}$ is the adjacent matrix of the network consisting of N nodes and $X \in \mathbb{R}^{N \times D}$ denotes the node feature matrix in which the node feature dimension is D. After obtaining the distribution, VGAE draws feature vectors, denoted by Z, from $\mathcal{N}(\mu, \sigma^2)$. Z is regarded as the embeddings of the network in low dimensional space. Finally, the network is constructed by

$$\hat{A} = sigmoid(ZZ^T). \tag{8.2}$$

The model is optimized with a cross-entropy loss to ensure the prediction accuracy and Kullback-Leibler divergence to keep the embeddings obeying Gaussian distribution. The total objective function is defined as

$$\begin{aligned} \mathcal{L}_{total} &= L_c - KL(Z, Z^n) \\ &= log(sigmoid(\hat{A}_{ij})) - \sum_i \sum_j Z_{ij} log(\frac{Z_{ij}}{Z^n_{ij}}), \end{aligned} \tag{8.3}$$

where Z_n is the noise drew from $\mathcal{N}(0, 1)$.

8.4   Experiments

8.5   Experiment Setup

We adopt the two methods introduced in Sect. 8.3 as well as HELP proposed in Chap. 2 to make friends and meal pals recommendations on Yelp dataset. Also, we would like to compare the performance of those methods with similarity index-based methods and node2vec. For node2vec, we set the embedding dimension to 128 and the hyper-parameters, p and q, are obtained through gird search over $\{0.50, 0.75, 1.00, 1.25, 1.50\}$. For VGAE, the number of units of the first GCN layer is 128 and the number of units of GCN_μ and GCN_σ are set as 64. The learning rate is set as 0.01 and the training iteration is taken as 200. And we set the number of neighbors of HELP as 35.

8.5.1   Friends Recommendation

Modeling the friendship on Yelp as a network, we regard the friends recommendation problem as a case of link prediction. We randomly remove a certain number of edges in the network which are considered as the potential friendship and the rest of the edges are used for training models. Tables 8.2 and 8.3 report the performance of the 9 link prediction methods on Yelp friendship network. All the methods

Table 8.2 The performance of friends recommendation with respect to AUC (The best results are in bold.)

	40%	50%	60%	70%	80%	90%
CN	0.6769	0.8252	0.7642	0.7962	0.8230	0.8406
SA	0.6728	0.7254	0.7568	0.7891	0.8174	0.8228
AC	0.6748	0.7238	0.7612	0.7886	0.8162	0.8312
HPI	0.7903	0.8067	0.8122	0.8239	0.8351	0.8363
HDI	0.8126	0.8142	0.8188	0.8238	0.8266	0.8212
SI	0.8185	0.8312	0.8346	0.8401	0.8435	0.8379
LHN	0.8393	0.8426	0.8430	0.8492	0.8523	0.8473
AA	0.6728	0.7260	0.7628	0.7980	0.8189	0.8438
RA	0.6719	0.7279	0.7650	0.7992	0.8256	0.8314
FM	0.6744	0.7261	0.7646	0.7955	0.8233	0.8445
LP	0.8052	0.8434	0.8639	0.8846	0.8990	0.9091
RF	0.8000	0.8446	0.8686	0.8867	0.9000	0.9015
N2V	0.8604	0.8732	0.8845	0.8965	0.9057	0.9078
VGAE	**0.8900**	**0.9262**	0.9144	0.9031	0.9062	0.8986
HELP	0.8801	0.9016	**0.9178**	**0.9289**	**0.9350**	**0.9396**

Table 8.3 The performance of friends recommendation with respect to AP (The best results are in bold.)

	40%	50%	60%	70%	80%	90%
CN	0.6750	0.7227	0.7610	0.7944	0.8209	0.8388
SA	0.6566	0.7044	0.7334	0.7668	0.7942	0.8102
JAC	0.6612	0.7047	0.7415	0.7689	0.7922	0.8144
HPI	0.7859	0.8059	0.8135	0.8277	0.8390	0.8446
HDI	0.8004	0.8045	0.8096	0.8157	0.8196	0.8131
SI	0.8220	0.8271	0.8305	0.8369	0.8409	0.8350
LHN	0.8412	0.8462	0.8446	0.8508	0.8554	0.8499
AA	0.6703	0.7243	0.7622	0.7982	0.8190	0.8443
RA	0.6707	0.7284	0.7663	0.8008	0.8246	0.8330
FM	0.6750	0.7273	0.7647	0.7955	0.8215	0.8436
LP	0.8041	0.8438	0.8652	0.8865	0.8994	0.9131
RF	0.7989	0.8452	0.8691	0.8863	0.9032	0.9037
N2V	0.8738	0.8835	0.8921	0.9032	0.9169	0.9141
VGAE	**0.9082**	**0.9464**	**0.9344**	0.9283	0.9213	0.9201
HELP	0.9020	0.9189	0.9321	**0.9410**	**0.9451**	**0.9497**

have relatively good performance when 90% edges are used for training. Not surprisingly, similarity index-based methods are not competitive as random walk-based approaches and deep learning models due to the lack of higher order network structures. Though RF assembles 11 similarity indices and improve the performance compared with single similarity index, it is still not as good as N2V, VGAE and HELP, especially when only a small number of edges are used for training. Further,

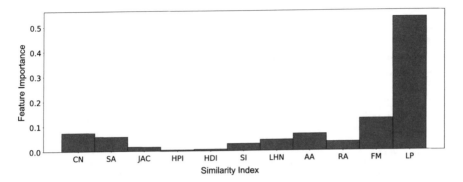

Fig. 8.4 Similarity index importance of G_f in RF model

the performance of all the methods improves as the amount of training edges gets larger. The improvement is particularly obvious in the similarity index-based methods while the performance of N2V, VGAE and HELP is relatively stable. Again, it proves the effectiveness of the threes methods. And we find that VGAE performs better than HELP when the percentage of training edges is smaller than 50% while HELP has better performance than VGAE when the amount of training edges increases. We argue that HELP needs more neighbor nodes to construct hyper-substructure network in sparser network. However, VGAE learns network structure features over the whole network and thus do not suffer the problem. When 90% of edges are used for training, the ability of network feature characterization enables HELP to outperforms other methods.

Besides the comparison of link prediction performance, we also investigate into the importance of the 11 similarity indices assembled by RF to see which similarity index matters the most. Figure 8.4 shows that LP is of the most significance in the prediction of links in G_f and it is also the most effective similarity index. And we find that FM which has poor performance but is the second major features in RF model, indicating that it is complementary to LP to certain extent.

8.5.2   Co-foraging Prediction

In this experiment, we investigate into the Yelp co-foraging network G_c to make meal pals recommendation. The same methods are adopted as we does in Sect. 8.5.1. As is shown in Tables 8.4 and 8.5, the similarity index-based methods are still not so effective but SA ranks the top in the link prediction of G_c. Yet, LP weights the most in the similarity indices assembly and SA takes the second place, which is recorded in Fig. 8.5. It is surprising to find that the performance of VGAE drops compared with other methods and RF performs the best under some circumstances. It might be due to that the change of network structures enhances the functionality

Table 8.4 The performance of co-foraging prediction with respect to AUC (The best results are in bold.)

	40%	50%	60%	70%	80%	90%
CN	0.6619	0.7266	0.7642	0.7756	0.8007	0.8636
SA	0.8970	0.9064	0.7568	0.9112	0.9053	0.9117
JAC	0.8679	0.8737	0.7612	0.8775	0.8725	0.8777
HPI	0.8586	0.8687	0.8842	0.8882	0.8942	0.9025
HDI	0.8526	0.8583	0.8571	0.8562	0.8504	0.8621
SI	0.8744	0.8776	0.7720	0.8763	0.8713	0.8821
LHN	0.8975	0.9011	0.9028	0.9034	0.9003	0.9069
AA	0.6649	0.7306	0.7628	0.7719	0.8041	0.8539
RA	0.6676	0.7282	0.7650	0.7764	0.8032	0.8392
FM	0.6682	0.7265	0.7705	0.8020	0.8020	0.8479
LP	0.8045	0.8529	0.8748	0.8950	0.8950	0.9078
RF	**0.9150**	**0.9254**	0.8686	0.9296	0.9379	0.9431
N2V	0.8880	0.9032	0.8845	0.9085	0.9221	0.9328
VGAE	0.7933	0.8242	0.9144	0.8557	0.8772	0.8954
HELP	0.9146	**0.9254**	**0.9178**	**0.9335**	**0.9412**	**0.9497**

Table 8.5 The performance of co-foraging prediction with respect to AP (The best results are in bold.)

	40%	50%	60%	70%	80%	90%
CN	0.6610	0.7249	0.7746	0.7995	0.8300	0.8626
SA	0.8954	0.9052	0.9107	0.9082	0.9183	0.9161
JAC	0.8550	0.8668	0.8699	0.8654	0.8747	0.8753
HPI	0.8384	0.8545	0.8735	0.8790	0.8860	0.8967
HDI	0.8526	0.8429	0.8458	0.8441	0.8387	0.8557
SI	0.8744	0.8679	0.8703	0.8693	0.8648	0.8794
LHN	0.8975	0.8978	0.9015	0.9001	0.8986	0.9053
AA	0.6628	0.7294	0.7709	0.8038	0.8249	0.8534
RA	0.6680	0.7284	0.7764	0.8031	0.8346	0.8414
FM	0.6682	0.7273	0.7714	0.8000	0.8347	0.8484
LP	0.8045	0.8521	0.8751	0.8943	0.9077	0.9086
RF	**0.9218**	**0.9306**	**0.9372**	**0.9461**	**0.9513**	**0.9508**
N2V	0.9022	0.9135	0.9148	0.9250	0.9328	0.9372
VGAE	0.8561	0.8787	0.9002	0.9142	0.9215	0.9265
HELP	0.8959	0.9066	0.9163	0.9273	0.9332	0.9376

of specific similarity indices like SA and LHN. The two indices weight more in the link prediction in G_c compared with their importance in the link prediction in G_f.

Assembling similarity indices, RF is not only effective but also explainable. Unlike VGAE and HELP which could not account for the prediction results, RF is able to interpret why the model makes such a decision.

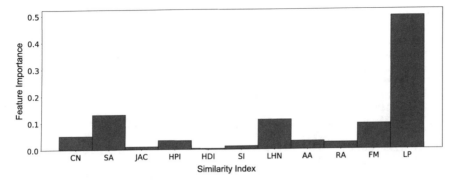

Fig. 8.5 Similarity index importance of G_c in RF model

8.6   Conclusion

In this chapter, we present the application of link prediction on Yelp friendship network and co-foraging network. The results show that network analysis has great potential in dealing with social commerce data. Friends and meal pals recommendation could improve the user experience and engagement of Yelp and thus leads to the promotion of revenues of the website and even the restaurants. Introducing network analysis techniques into recommendation on websites like Yelp has broad technical prospect and of high commercial value. In this basic application, we mainly use the network structures to make link prediction. Many other attributes like the reviews and users' profile s could also be integrated into the methods we present in Sect. 8.3, which may lead to better performance. And modeling Yelp as a heterogeneous network or a temporal network may be a more proper way to analyze such daedal data, which could be further studied in future works.

References

1. Yang, S.B., Hlee, S., Lee, J., Koo, C.: An empirical examination of online restaurantreviews on yelp.com: A dual coding theory perspective. International Journal of Contemporary Hospitality Management, Emerald Publishing Limited (2017)
2. Huang, J., Rogers, S., Joo, E.: Improving restaurants by extracting subtopics from yelp reviews. In: iConference 2014 (Social Media Expo) (2014)
3. Cervellini, P., Menezes, A.G., Mago, V.K.: Finding trendsetters on yelp dataset. In: 2016 IEEE Symposium Series on Computational Intelligence (SSCI), pp. 1–7. IEEE, Piscataway (2016)
4. Sáez-Trumper, D., et al.: Finding relevant people in online social networks. Ph.D. Thesis, Universitat Pompeu Fabra (2014)
5. Xuan, Q., Zhou, M., Zhang, Z.Y., Fu, C., Xiang, Y., Wu, Z., Filkov, V.: Modern food foraging patterns: geography and cuisine choices of restaurant patrons on yelp. IEEE Trans. Comput. Soc. Syst. **5**(2), 508–517 (2018)

6. Fu, C., Zhao, M., Fan, L., Chen, X., Chen, J., Wu, Z., Xia, Y., Xuan, Q.: Link weight prediction using supervised learning methods and its application to yelp layered network. IEEE Trans. Knowl. Data Eng. **30**(8), 1507–1518 (2018)

7. Yu, X., Ren, X., Sun, Y., Gu, Q., Sturt, B., Khandelwal, U., Norick, B., Han, J.: Personalized entity recommendation: A heterogeneous information network approach. In: Proceedings of the 7th ACM International Conference on Web Search and Data Mining, pp. 283–292 (2014)

8. Lü, L., Zhou, T.: Link prediction in complex networks: a survey. Phys. A Stat. Mech. Appl. **390**(6), 1150–1170 (2011)

9. Cui, P., Wang, X., Pei, J., Zhu, W.: A survey on network embedding. IEEE Trans. Knowl. Data Eng. **31**(5), 833–852 (2018)

10. Zhang, M., Chen, Y.: Link prediction based on graph neural networks. In: Proceedings of 32nd NeurIPS, pp. 5171–5181 (2018)

11. Salha, G., Limnios, S., Hennequin, R., Tran, V.A., Vazirgiannis, M.: Gravity-inspired graph autoencoders for directed link prediction. In: Proceedings of the 28th CIKM, pp. 589–598 (2019)

12. Zhang, J., Zheng, J., Chen, J., Xuan, Q.: Hyper-substructure enhanced link predictor. In: Proceedings of the 29th ACM International Conference on Information & Knowledge Management, pp. 2305–2308 (2020)

13. Kipf, T.N., Welling, M.: Variational graph auto-encoders. Preprint. arXiv:1611.07308 (2016)

Chapter 9
Graph Convolutional Recurrent Neural Networks: A Deep Learning Framework for Traffic Prediction

Dongwei Xu, Hongwei Dai, and Qi Xuan

Abstract Road traffic state prediction is a challenging task for urban traffic control and guidance due to the complicated spatial dependencies on the roadway network and the time-varying traffic flow data. In this work, a novel traffic flow prediction method named the graph convolutional recurrent neural network (GCRNN) is proposed to tackle this challenge. First, we address the problem on a graph and build the model with graph embedding techniques. Second, the proposed model employs the GCN model to learn the interactions of the roadways to capture the spatial dependence and uses the long short-term memory (LSTM) neural network (NN) to learn dynamic changes of traffic data to capture temporal dependence. An experiment is conducted on a Hangzhou transportation network with several typical intersections under the Sydney coordinated adaptive traffic system (SCATS), the results of which indicate that our model yields excellent performance in terms of different prediction error measures.

9.1 Background

Accurate traffic prediction is a key part of an advanced traffic management system because traffic conditions have direct effects on the sustainable development of modern cities[1, 2]. For example, traffic congestion is usually resulted from the gathering of people and influences the outdoor activities. Therefore, high-accuracy traffic prediction can help control a serious of traffic accidents in advance[3, 4]. Thus, the Sydney Coordinated Adaptive Traffic System (SCATS) is widely used in the world, and the efficient prediction of traffic flow has received much attention.

Traffic data contain spatial location and temporal dynamics information. Since a road's traffic data is influenced by its upstream and downstream roadways, we

D. Xu, H. Dai, Q. Xuan (✉)
Institute of Cyberspace Security, Zhejiang University of Technology, Hangzhou, China

College of Information Engineering, Zhejiang University of Technology, Hangzhou, China
e-mail: xuanqi@zjut.edu.cn

© The Author(s), under exclusive license to Springer Nature Singapore Pte Ltd. 2021   189
Q. Xuan et al. (eds.), *Graph Data Mining*, Big Data Management,
https://doi.org/10.1007/978-981-16-2609-8_9

need to consider spatial factor into traffic prediction. In addition, traffic data at a certain time is dependent on the traffic state during the previous and the latter time interval. Thus traffic data have spatiotemporal correlation, and traffic prediction has spatiotemporal characteristics.

Traffic state prediction has been studied for decades. There are quite many approaches for traffic state prediction. Statistical models, such as autoregressive integrate moving average (ARIMA) model[5], Kalman filter (KF)[6], and their variants[7–9], have been applied to predict future traffic flow based on previously observed values using time-series analysis.

However, the traditional statistical models typically depend on the assumption of stationary, which cannot fit the nonstationary and uncertainty characteristics of traffic data. In comparison with the statistical models, machine learning models are free of assumptions on traffic data distribution and only require enough traffic data to learn the regularity automatically. The common machine learning models include Bayesian networks model[10], k-nearest neighbor (KNN) algorithm[11–15], and support vector machine (SVM) model[16, 17], all of which yield good prediction results.

Recently, with the rapid development of deep learning, the deep neural network models have been applied widely and successfully to predict traffic flow, such as deep belief networks (DBNs)[18], stacked autoencoders (SAEs)[19], convolutional neural networks (CNNs)[20, 21], and recurrent neural networks (RNNs)[22–24], which have been proven to be efficient in traffic flow prediction. Till date, CNNs are effective in recognizing spatial patterns. The RNNs and its variants (long short-term memory (LSTM)[22] and the gated recurrent unit (GRU)[25]) can capture nonlinear traffic dynamics effectively and learn temporal dependence well.

However, there are some limitations in the existing traffic prediction methods for urban road networks. (1) In the process of traffic prediction, temporal-spatial correlation is an important factor that has to be considered. Temporal correlations[26] denote the correlations between the current traffic state and past traffic state with a temporal span, whereas spatial correlations mean the correlations of the selected road segment and those of its upstream and downstream road segments at the same time interval. Thus, we must take full advantage of spatial features. (2) For traffic state prediction, traffic flow data are obtained from the loop detectors. The loop detector is a device that identifies the passing or presence of a vehicle. However, road traffic state prediction for urban road networks needs specific traffic flow data. In the SCATS, the obtainable information is the traffic flow at the entrance[27] of each roadway in the intersections, rather than the traffic flows of the road segments, which makes the traditional traffic flow prediction of an urban road network difficult to achieve.

In this chapter, to overcome the shortage of methods mentioned above, we represent the traffic network as a graph and take the topology of the road network into account. This paper proposes a novel scheme, named the graph convolutional recurrent neural network (GCRNN), to model the dynamics of the traffic flow and make the prediction more accurate. The contributions of this work mainly include the following points:

- A complex network of the urban road network is constructed based on a directed weighted network, which can make full use of the data collected from the loop detectors of the SCATS and obtain more useful information of the intersections.
- GCRNN uses the graph convolutional network(GCN) model to learn structural regularities presented within convolution operation and builds robust representations which are suitable for capture spatial features.
- The LSTM neural network is used to learn dynamic temporal dependencies presented in traffic data based on vector representations we learned.
- We evaluate our scheme using the traffic data of Hangzhou. The experimental results demonstrate the advantages of the new scheme that we proposed compared with other baselines.

9.2 Related Work

9.2.1 Graph Analysis

Graphs have been used to represent information in various areas including biology[28], social sciences[29], and linguistics[30]. Graph analysis are also applied to traffic domain since traffic networks can be represented by nodes and edges. Graph-structured data appear frequently in the transportation field. In the road traffic complex network, the most recent state and the next state of the road traffic network could be constrained by a certain regulation. Graph embedding can capture abundant significant information from the complex network. In recent years, the graph convolutional network (GCN)[31] is popular in traffic field. The GCN model can tackle the traffic prediction problem because it can learn irregular matrix. The GCN model uses the convolution operation in the Fourier domain to capture the topological structure of traffic network.

9.2.2 Traffic State Prediction

Traffic state prediction is a typical time-series prediction problem. The aim of traffic forecasting is to learn a function $h(\cdot)$ from previous traffic observations to predict future traffic state. We define a traffic state vector $X_t \in \mathbb{R}^N$ to represent the traffic state information of N roadways in the road network at the time step t. Given the historical time series until time step t, the predicted traffic state can be defined as:

$$[X_{t-T}, \ldots, X_{t-1}, X_t] \xrightarrow{h(\cdot)} \widetilde{X}_{t+1}, \tag{9.1}$$

where $\widetilde{X}_{t+1}$ represents the predicted value at the next time step, and T is the length of historical time series.

In network-wide traffic prediction, road segments have a series of characteristics that can be carried by the node in the graph, which often include the average speed, occupancy, vehicle volume, and so on. Under the SCATS system, only the entrance roadways have traffic state data. So, a directed road network can be described as follows:

$$G = (V, E, A). \tag{9.2}$$

where, $V = \{v_1, v_2, \ldots, v_N\}$ is a finite set of nodes, $N = |V|$ is the number of nodes; $E = \{e_{i,j} | i, j \in N\}$ is the set of edges, $e_{i,j} \neq e_{j,i}$. In this chapter, a node represents a roadway, and an edge represents that two roadways are connected. $A \in \mathbb{R}^{N \times N}$ is an adjacent matrix of the traffic network G. The elements in the adjacent matrix represent the connectedness of nodes. $A_{i,j} = 1$ denotes that there is a roadway along the driving direction connecting node i and node j, and $A_{i,j} = 0$ otherwise. Specifically, $A_{i,i} = 0$.

Thus, the traffic state prediction problem on the road network G can be formulated as

$$[X_{t-T}, \ldots, X_{t-1}, X_t; G] \xrightarrow{h(\cdot)} \widetilde{X}_{t+1}. \tag{9.3}$$

9.3 Model

To capture the spatial and temporal dependencies from traffic data jointly, we propose a graph convolutional recurrent neural network (GCRNN) model that combines spatial dependency modeling and temporal dynamics modeling. The architecture of the proposed model is shown in Fig. 9.1.

In the figure, we take the historical time series $\{X_{t-n}, \ldots, X_{t-1}, X_t\}$ and the adjacent matrix A as input of GCRNN. GCRNN first utilizes GCN to capture spatial features of network structure from the graph-structural traffic data of each time step, which can be expressed as $f(X, A)$. $f(X, A)$ denotes the output and d is the number of hidden layer unit of GCN. Second, the time series with obtained features $[f(X_{t-n}, A), \ldots, f(X_{t-1}, A), f(X_t, A)]$ are fed into the LSTM model to extract the temporal features. Finally, we get predictions with prediction length T through a fully connected layer.

9.3.1 Graph Convolutional Network

In the proposed method, we use GCN to capture the spatial features of the road network. The architecture of GCN is shown in Fig. 9.2. Schematic depiction of regression task on graph with T input channels and F features map in output layer.

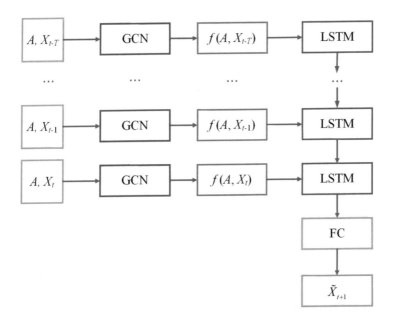

Fig. 9.1 The architecture of the proposed model

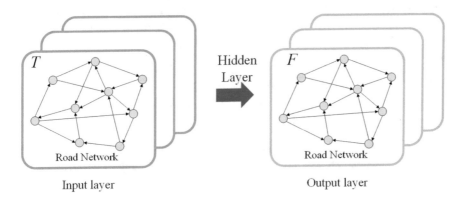

Fig. 9.2 The architecture of GCN

GCN model constructs a filter in the Fourier domain. The filter operates on each node and its first-order neighborhood to extract the spatial features.

A multi-layer GCN model with the layer-wise propagation rule is shown as follow:

$$H^{(l+1)} = \mathscr{F}\left(\tilde{D}^{-1/2}\tilde{A}\tilde{D}^{-1/2}H^{(l)}W^{(l)}\right), \qquad (9.4)$$

where $\tilde{A} = A + I_N$ is the adjacency matrix with added self-connections; $I_N \in \mathbb{R}^{N \times N}$ is is an identity matrix; $\tilde{D} \in \mathbb{R}^{N \times N}$ is the degree matrix, $\tilde{D} = \sum_j \tilde{A}_{ij}$; $H^{(l)}$ is the output of l layer, $H^{(0)} = X \in \mathbb{R}^{N \times T}$; $W^{(l)}$ is the trainable parameters of that layer; and $\mathscr{F}(\cdot)$ is the activation function such as ReLU.

For example, a 2-layer GCN model can be expressed as:

$$f(X, A) = \mathscr{F}\left(\tilde{D}^{-1/2} \tilde{A} \tilde{D}^{-1/2} \text{ReLU}\left(\tilde{D}^{-1/2} \tilde{A} \tilde{D}^{-1/2} X W^{(0)}\right) W^{(1)}\right). \qquad (9.5)$$

Here, $W^{(0)} \in \mathbb{R}^{T \times H}$ is an input-to-hidden weight matrix; $W^{(1)} \in \mathbb{R}^{H \times F}$ is a hidden-to-output weight matrix; and H is the hidden units of hidden layer.

In this chapter, we introduce a spectral convolution on graphs defined as the multiplication of a signal X with a filter $g_\theta = \text{diag}(\theta)$ parameterized by $\theta \in \mathbb{R}^N$ in the Fourier domain[32]. The $\text{diag}(\theta)$ is the diagonalized matrix given θ. The spectral graph convolution operation can be defined as

$$g_\theta *_\mathscr{G} X = U g_\theta U^T X = U \text{diag}(\theta) U^T X, \qquad (9.6)$$

where $*_\mathscr{G}$ is the spectral graph convolution operator; U is the matrix of eigenvectors of the normalized graph Laplacian $L = I_N - D^{-\frac{1}{2}} A D^{-\frac{1}{2}} = U \Lambda U^T$, with a diagonal matrix of its eigenvalues $\Lambda = \text{diag}([\lambda_0, \lambda_1, \ldots, \lambda_N]) \in \mathbb{R}^{N \times N}$; and $U^T X$ being the graph Fourier transform of X. We can consider g_θ as a learnable convolutional kernel weight. However, the eigenvalue decomposition of the Laplacian matrix is computationally expensive. Therefore, Chebyshev polynomials apply to solve this problem:

$$g_\theta(\Lambda) = \sum_{k=0}^{K} \theta_k T_k\left(\tilde{\Lambda}\right) X, \qquad (9.7)$$

where $\tilde{\Lambda} = \frac{2}{\lambda_{max}} \Lambda - I_N$, λ_{max} denotes the largest eigenvalue of L, and $\theta_k \in \mathbb{R}^{k \times N}$ is trainable parameter. The recursive definition of the Chebyshev polynomial is denoted as $T_k(x) = 2x T_{k-1}(x) - T_{k-2}(x)$, where $T_k(\tilde{L}) \in \mathbb{R}^{N \times N}$ is the Chebyshev polynomial of order k, $T_0(x) = 1$, and $T_1(x) = x$.

Then the definition of a convolution of a signal X with a filter g_θ is shown as:

$$g_\theta *_\mathscr{G} X = \sum_{k=0}^{K} \theta_k T_k\left(\tilde{L}\right) X, \qquad (9.8)$$

where $\tilde{L} = \frac{2}{\lambda_{max}} L - I_N$. Approximate expansion of Chebyshev polynomials can solve this formulation corresponding to extract information of surrounding 0 to $(K-1)^{\text{th}}$ -order neighbors centered on each node in the graph by the convolution kernel g_θ.

To reduce the computational complexity, a two-layer GCN model can be defined by stacking multiple localized graph convolutional layers with only first-order approximation. We can assume $\lambda_{max} \approx 2$ in scale during training, Eq. 9.5 can be simplified to:

$$g_\theta *_\mathcal{G} X \approx \theta_0 x + \theta_1 (L - I_N) X = \theta_0 x - \theta_1 \left(D^{-\frac{1}{2}} A D^{-\frac{1}{2}} \right) X, \qquad (9.9)$$

where θ_0, θ_1 are two shared kernel parameters. In order to constrain the number of parameters further to address overfitting and to minimize the number of operations per layer, the graph convolution can be expressed as:

$$g_\theta *_\mathcal{G} X = \theta \left(I_N + D^{-\frac{1}{2}} A D^{-\frac{1}{2}} \right) X, \qquad (9.10)$$

where θ_0, θ_1 are replaced by a single parameter θ by letting $\theta_0 = \theta_1 = \theta$.

Thus, the output $f(X, A)$ of the GCN model on the spectral convolution on graphs can be defined as:

$$f(X, A) = \mathcal{F} \left(\sum_{k=0}^{K} \theta_k T_k \left(\tilde{L} \right) X \right). \qquad (9.11)$$

9.3.2   Long Short-Term Memory

With the ability to capture temporal dependencies, the recurrent NN is usually applied in time-series analysis. RNNs were initially used for language models, due to its ability to memorize long-term dependencies. When time lags increase, gradients of RNNs may vanish through unfolding RNNs into very deep feed forward neural networks. To address the gradient vanishing problem, a representative variant of RNNs, like LSTM[33], is designed to give the memory cells ability to determine when to forget certain information, thus determining the optimal time lags.

The LSTM neural network has a complex structure named LSTM cell in its hidden layer. The architecture of the LSTM model is shown in Fig. 9.3. A typical LSTM cell is composed mainly of three gates, namely, input gate, forget gate, and output gate, which control the information flow through the cell and the neural network. Input gate takes a new input point from outside and processes newly coming data. Memory cell gets input from the output of the LSTM cell in the last iteration. Forget gate decides when to forget the output results and thus selects the optimal time lag for the input sequence. At time t, the input historical traffic state is $X_t \in \mathbb{R}^N$, the hidden layer output is h_t and its former output is h_{t-1}, the cell input state is $\tilde{c}_t$, and the output state is c_t and its former state is c_{t-1}. The states of the three gates are i_t, f_t, and o_t. The input gate can determine how many new candidate cell can be added to the cell state.

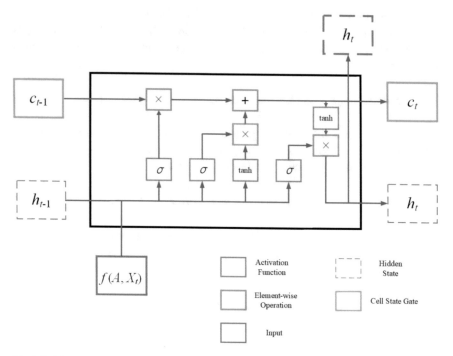

Fig. 9.3 The architecture of LSTM

The structure of the LSTM cell indicates that both c_t and h_t are transmitted to the next neural network. The calculation procedures of c_t and h_t are shown as follows:

$$i_t = \sigma \left(W_i X_t + U_i h_{t-1} + b_i \right), \tag{9.12}$$

$$f_t = \sigma \left(W_f X_t + U_f h_{t-1} + b_f \right), \tag{9.13}$$

$$\tilde{c}_t = \tanh \left(W_c X_t + U_c h_{t-1} + b_c \right), \tag{9.14}$$

$$c_t = f_t \odot c_{t-1} + i_t \odot \tilde{c}_t, \tag{9.15}$$

$$o_t = \sigma \left(W_o X_t + U_o h_{t-1} + b_o \right), \tag{9.16}$$

$$h_t = o_t \odot \tanh \left(c_t \right), \tag{9.17}$$

where W_i, W_f and W_o are weight matrices of the input gate, the forget gate and the output gate, respectively; U_i, U_f and U_o are weight matrices for the preceding hidden state; b_i, b_f and b_o are their bias vectors; $\sigma (\cdot)$ denotes the activation function of the gate; and $\odot$ stands for the scalar product of two vectors or matrices.

9.3.3 Graph Convolutional Recurrent Neural Network

In summary, the GCRNN model can deal with the spatial dependence and temporal dynamics. First, GCN is used to capture the topological structure of the road network to receive the spatial dependence. The input historical traffic state X_t is fed into the GCN function, and the output $f(X_t, A) \in \mathbb{R}^K$ is shown as:

$$f(X_t, A) = \mathscr{F}\left(\sum_{k=0}^{K} \theta_k T_k\left(\tilde{L}\right) X_t\right), \tag{9.18}$$

Second, the LSTM is used to learn the dynamic change from time series to obtain the temporal dependencies. The calculation process is shown below:

$$i_t = \sigma\left(W_i \cdot f(X_t, A) + U_i h_{t-1} + b_i\right), \tag{9.19}$$

where $f(X_t, A)$ is the input of input gate $i_t \in [0, 1]$; $W_i \in \mathbb{R}^{K \times d}$ and $b_i \in \mathbb{R}^d$ are weight matrix and bias vector, respectively; and U_i denotes the weight matrix for the preceding hidden state.

$$f_t = \sigma\left(W_f \cdot f(X_t, A) + U_f h_{t-1} + b_f\right), \tag{9.20}$$

where $f(X_t, A)$ is the input of forget gate $f_t \in [0, 1]$; $W_f \in \mathbb{R}^{K \times d}$ and $b_f \in \mathbb{R}^d$ are weight matrix and bias vector of forget gate, respectively; and U_f denotes the weight matrix for the preceding hidden state. The next step is to update the cell state:

$$\tilde{c}_t = \tanh\left(W_c \cdot f(X_t, A) + U_c h_{t-1} + b_c\right), \tag{9.21}$$

where $f(X_t, A)$ is the input of new cell state $\tilde{c}_t$; $W_c \in \mathbb{R}^{K \times d}$ and $b_c \in \mathbb{R}^d$ are weight matrix and bias vector of forget gate; and U_c denotes the weight matrix for the preceding hidden state. The cell state c_t can be updated by the forget gate $f_t \in [0, 1]$, the input gate $i_t \in [0, 1]$ and the candidate cell $\tilde{c}_t \in [-1, 1]$.

$$c_t = f_t \odot c_{t-1} + i_t \odot \tilde{c}_t, \tag{9.22}$$

The updated cell layer information can not only have long-term information, but also selectively filter out some useless information. The output gate o_t can decide what information can be the output:

$$o_t = \sigma \left(W_o \cdot f \left(X_t, A \right) + U_o h_{t-1} + b_o \right), \tag{9.23}$$

$$h_t = o_t \odot \tanh \left(c_t \right), \tag{9.24}$$

where $f(X_t, A)$ is the input of output gate at time t; $W_o \in \mathbb{R}^{K \times d}$ and $b_o \in \mathbb{R}^d$ are weight matrix and bias vector of forget gate; and U_o denotes the weight matrix for the preceding hidden state, and $h_t \in \mathbb{R}^d$ is the output of the LSTM model which contains the spatiotemporal information at time t.

Finally, the output vectors of the GCRNN are used as the input into the fully connected layer to realize the traffic prediction as shown below:

$$\tilde{X}_{t+1} = \sigma \left(W_{fc} h_t + b_{fc} \right), \tag{9.25}$$

where $\tilde{X}_{t+1}$ are the prediction results corresponding to N nodes; $W_{fc} \in \mathbb{R}^{d \times N}$ and $b_{fc} \in \mathbb{R}^N$ are weight matrix and bias vector of fully connected network for linear transformation, respectively; and $\sigma(\cdot)$ denotes the sigmoid function in this chapter.

In the training process, the goal is to minimize the error between the true values and the predicted values. The loss during the training process is defined as:

$$Loss = L \left(Y_t, \tilde{Y}_t \right) = \sum_{t=1}^{M} \left(Y_t - \tilde{Y}_t \right)^2, \tag{9.26}$$

where $L(\cdot)$ is a loss function to calculate the error between Y_t and $\tilde{Y}_t$; Y_t and $\tilde{Y}_t$ are represent the true traffic data and the predicted data, respectively; $Y_t = X_{t+1}$, $\tilde{Y}_t = \tilde{X}_{t+1}$; and M is the number of time samples. In this chapter, we use the mean square error (MSE) as the loss function.

9.4 Experiment

9.4.1 Dataset

In this study, the dataset was obtained from the SCATS of 17 intersections in Hangzhou. The map of these intersections (red) is shown in the left of Fig. 9.4. We numbered the road segments of the selected intersections. The spatial topology of the road network is shown in the right of Fig. 9.4. The traffic volume data are extracted from June 1 to 30, 2017 for this experiment. The collected dataset is divided into the training set (80% of the overall data) and the testing set (the remaining data). The time horizon of records is 00:00–24:00, and the traffic state data collection interval is 15 min. So a roadway has 96 data points per day.

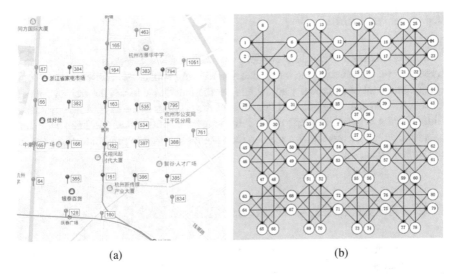

(a)   (b)

Fig. 9.4 Left: The map of the selected intersections (red). Right: The spatial topology of the road network based on the selected area

9.4.2 Baselines

In this subsection, the proposed model is compared with the following models: (1) KNN: k-nearest neighbors, (2) FNN: feed forward neural network, (3) LSTM: long short-term memory recurrent neural network, and (4) GCN: graph convolutional network. FNN and LSTM are implemented based on keras, GCN and GCRNN are built by tensorflow, and KNN is implemented with scikit-learn.

9.4.3 Evaluation

In this experiment, the performances of the proposed and the compared models are evaluated by two commonly used metrics in traffic prediction, including (1) root mean squared error (RMSE) and (2) mean absolute error (MAE).

$$\text{RMSE} = \left(\frac{1}{M}\sum_{i=1}^{M}\left(Y_t - \tilde{Y}_t\right)\right)^{\frac{1}{2}},\tag{9.27}$$

$$\text{MAE} = \frac{1}{M}\sum_{i=1}^{M}\left|Y_t - \tilde{Y}_t\right|.\tag{9.28}$$

9.4.4 Evaluation

The aforementioned baseline models and the GCRNN model have the same hyperparameters, such as the batch size, the training epoch, and the learning rate. In this experiment, we manually adjust and set the batch size to 64, the training epoch to 100, and the learning rate to 0.001. We set the number of the LSTM layer to 2 and hidden units d in each layer to 256. All the tests use 60 min as the historical time window, which means that 4 observed data points are used to predict traffic states in the next 15 min.

9.4.5 Results of Experiments and Analyses

K is a very important hyperparameter because the K-order Chebyshev polynomial approximates the graph convolution. Thus, the GCRNN model can utilize the information of the node which are at the largest K hop from the central node. When $K = 1$, only the information of the central node is considered. When $K = 2$, the relation of information of the first-order neighbor nodes of the central node are considered. When $K = 3$, the information on the first-order neighbor nodes and second neighbor nodes of the central node are additionally considered. The larger K is that the more additional structure information from the neighbor nodes of the central node can be considered, but the computation complexity is greatly increased. Normally, K is set to 3.

We predict all the roadways and give the comparison between our model and baselines with metrics RMSE and MAE as shown in Table 9.1.

In our experiment, different number of hidden units d in GCRNN may greatly affect the prediction performance. To choose the suitable hyperparameter d, we do experiments with different number of hidden unit and select the optimal value by comparing the predictions. Generally, the number of hidden units d is selected from $\{16, 32, 64, 128, 256, 512\}$. The result is shown in Table 9.2.

As shown in Table 9.2, the errors reach a minimum when the number is 256. In practice, the larger N is, the more hidden units we need to use in the model. When the number of hidden units is more than 256, the prediction precision decreases. This is mainly because when the hidden units are larger than a certain degree, the

Table 9.1 The prediction results of the GCRNN model and other baseline methods

Model	RMSE	MAE
KNN	15.76	10.20
FNN	14.76	7.05
LSTM	13.73	6.70
GCN	13.76	6.98
GCRNN	12.30	6.46

Table 9.2 The prediction
results of the GCRNN model
under different hidden units d

d	RMSE	MAE
16	14.80	6.96
32	14.70	6.89
64	13.89	6.46
128	12.63	6.32
256	12.30	6.13
512	12.38	6.21

model complexity is increased and as a result, overfitting occurs. Thus, we set the number of hidden units to 256 in our experiments.

The proposed method outperforms other models with different metrics. The traditional machine learning algorithm KNN shows the weakness when compared with other deep learning methods. This is mainly due to the KNN is difficult to handle complex, nonstationary time series data. Deep learning methods generally acquires better prediction results than the traditional machine learning models. FNN does not perform well on forecasting spatial-temporal sequence. The LSTM model outperforms the KNN method and the FNN method because it can capture the temporal correlations of traffic flow data. However, the LSTM only considers the temporal features and ignore the spatial features. The GCN model have a better performance than the LSTM model, indicating that the spatial features is great important in traffic flow prediction. However, the GCN only considers the spatial features and ignore the temporal features. Because network-wide traffic prediction has the spatial-temporal correlations, the GCRNN model has the lowest RMSE error and MAE error compare the LSTM method and the GCN method. The GCRNN model achieve best results, verifying that graph convolution is a reasonable method to learn the interaction between the roadways in traffic prediction problem.

Figure 9.5 visualizes the predicted traffic flow series and the truth of two locations selected from the dataset. We can clearly obverse that the GCRNN model generates smooth prediction and is usually better at predicting the start and the end of peak hours. Due to the stronger influence between the intersections in the morning peak and evening rush hours, the increasing volume on a roadway could cause traffic congestion in other roadways. GCRNN can capture the spatial relationships of the traffic network. It indicates that our proposed method prefers the peak hours with the big data and the traffic spatial correlation is not significant in low volume condition. These results demonstrate the superiority of our GCRNN model over other baselines, which show promising potential in road traffic state prediction.

9.5 Conclusion

In this study, a GCRNN approach is proposed to tackle the challenge of traffic flow prediction in the SCATS. We represent the traffic road network as a graph and employ graph embedding to capture the topological structure of the graph to receive

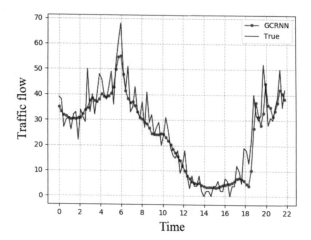

(a) Comparison between the true values and the predicted values

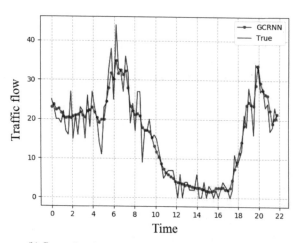

(b) Comparison between the true values and the predicted values

Fig. 9.5 Traffic flow prediction visualization on two randomly selected days

the spatial features, which cannot be accomplished by previously utilized methods. The LSTM model can get a stable performance in time-series traffic forecasting. The experimental results demonstrate that the GCRNN model outperforms other baseline models.

In the future, we will consider the other factors as features, such as weather, events and periodicity. In addition, more traffic flow theories also can be integrated into the model.

References

1. Xu, D., Dong, H., Jia, L., Qin, Y.: Virtual speed sensors based algorithm for expressway traffic state estimation. Sci. Chin. Technol. Sci. **55**(5), 1381–1390 (2012)
2. Xu, D.-w., Dong, H.-h., Jia, L.-m., Tian, Y.: Road traffic states estimation algorithm based on matching of regional traffic attracters. J. Central South Univ. **21**(5), 2100–2107 (2014)
3. Zheng, Y., Capra, L., Wolfson, O., Yang, H.: Urban computing: concepts, methodologies, and applications. ACM Trans. Intel. Syst. Technol. **5**(3), 38 (2014)
4. Xu, D., Dong, H., Li, H., Jia, L., Feng, Y.: The estimation of road traffic states based on compressive sensing. Transportmetrica B Trans. Dyn. **3**(2), 131–152 (2015)
5. Hamed, M.M., Al-Masaeid, H.R., Bani Said, Z.M.: Short-term prediction of traffic volume in urban arterials. J. Transp. Eng. **121**(3), 249–254 (1995)
6. Okutani, I., Stephanedes, Y.J.: Dynamic prediction of traffic volume through kalman filtering theory. Transp. Res. B Methodol. **18**(1), 1–11 (1984)
7. Williams, B.M., Hoel, L.A.: Modeling and forecasting vehicular traffic flow as a seasonal arima process: theoretical basis and empirical results. J. Transp. Eng. **129**(6), 664–672 (2003)
8. Guo, J., Huang, W., Williams, B.M.: Adaptive kalman filter approach for stochastic short-term traffic flow rate prediction and uncertainty quantification. Trans. Res. C Emer. Technol. **43**, 50–64 (2014)
9. Xu, D.-w., Wang, Y.-d., Jia, L.-m., Qin, Y., Dong, H.-h.: Real-time road traffic state prediction based on arima and kalman filter. Front. Inf. Technol. Elect. Eng. **18**(2), 287–302 (2017)
10. Fusco, G., Colombaroni, C., Isaenko, N.: Short-term speed predictions exploiting big data on large urban road networks. Trans. Res. C Emer. Technol. **73**, 183–201 (2016)
11. Zhang, Y., Zhang, Y., Haghani, A.: A hybrid short-term traffic flow forecasting method based on spectral analysis and statistical volatility model. Trans. Res. C Emer. Technol. **43**, 65–78 (2014)
12. Cai, P., Wang, Y., Lu, G., Chen, P., Ding, C., Sun, J.: A spatiotemporal correlative k-nearest neighbor model for short-term traffic multistep forecasting. Trans. Res. C Emer. Technol. **62**, 21–34 (2016)
13. Xu, D., Wang, Y., Jia, L., Zhang, G., Guo, H.F.: Real-time road traffic states estimation based on kernel-KNN matching of road traffic spatial characteristics. Journal of Central South University **23**(9), 2453–2464 (2016)
14. Xu, D., Wang, Y., Peng, P., Beilun, S., Deng, Z., Guo, H.: Real-time road traffic state prediction based on kernel-KNN. Transportmetrica A Trans. Sci. 1–15 (2018)
15. Lin, L., Li, Y., Sadek, A.: A k nearest neighbor based local linear wavelet neural network model for on-line short-term traffic volume prediction. Procedia. Soc. Behav. Sci. **96**, 2066–2077 (2013)
16. Lin, L., Wang, Q., Sadek, A.W.: Short-term forecasting of traffic volume: evaluating models based on multiple data sets and data diagnosis measures. Transp. Res. Rec. **2392**(1), 40–47 (2013)
17. Castro-Neto, M., Jeong, Y.-S., Jeong, M.-K., Han, L.D.: Online-SVR for short-term traffic flow prediction under typical and atypical traffic conditions. Exp. Syst. Appl. **36**(3), 6164–6173 (2009)
18. Huang, W., Song, G., Hong, H., Xie, K.: Deep architecture for traffic flow prediction: deep belief networks with multitask learning. IEEE Trans. Intel. Transp. Syst. **15**(5), 2191–2201 (2014)
19. Lv, Y., Duan, Y., Kang, W., Li, Z., Wang, F.Y.: Traffic flow prediction with big data: a deep learning approach. IEEE Trans. Intel. Transp. Syst. **16**(2), 865–873 (2015)
20. Ma, X., Dai, Z., He, Z., Ma, J., Wang, Y., Wang, Y.: Learning traffic as images: a deep convolutional neural network for large-scale transportation network speed prediction. Sensors **17**(4), 818 (2017)
21. Liu, Q., Wang, B., Zhu, Y.: Short-term traffic speed forecasting based on attention convolutional neural network for arterials. Comput. Aided Civ. Inf. Eng. **33**(11), 999–1016 (2018)

22. Ma, X., Tao, Z., Wang, Y., Yu, H., Wang, Y.: Long short-term memory neural network for traffic speed prediction using remote microwave sensor data. Trans. Res. C Emer. Technol. **54**, 187–197 (2015)

23. Van Lint, J.W.C., Hoogendoorn, S.P., van Zuylen, H.J.: Freeway travel time prediction with state-space neural networks: modeling state-space dynamics with recurrent neural networks. Transp. Res. Rec. **1811**(1), 30–39 (2002)

24. Cui, Z., Ke, R., Wang, Y.: Deep bidirectional and unidirectional LSTM recurrent neural network for network-wide traffic speed prediction. arXiv preprint arXiv:1801.02143 (2018)

25. Chung, J., Gulcehre, C., Cho, K.H., Bengio, Y.: Empirical evaluation of gated recurrent neural networks on sequence modeling. Preprint. arXiv:1412.3555 (2014)

26. Wang, Y., Xu, D., Lu, Y., Shen, J., Zhang, G.: Compression algorithm of road traffic data in time series based on temporal correlation. IET Intel. Trans. Syst. **12**(3), 177–185 (2018)

27. Xu, D.W., Wang, Y.D., Li, H.J., Qin, Y., Jia, L.M.: The measurement of road traffic states under high data loss rate. Measurement **69**, 134–145 (2015)

28. Theocharidis, A., Van Dongen, S., Enright, A.J., Freeman, T.C.: Network visualization and analysis of gene expression data using biolayout express 3d. Nat. Prot. **4**(10), 1535 (2009)

29. Freeman, L.: Visualizing social networks. Soc. Netw. Data Anal. **6**(4), 411–429 (2000)

30. Cancho, R.F.I., Solé, R.V.: The small world of human language. Proc. R. Soc. Lond. Ser. B Biol. Sci. **268**(1482), 2261–2265 (2001)

31. Kipf, T.N., Welling, M.: Semi-supervised classification with graph convolutional networks. arXiv preprint arXiv:1609.02907 (2016)

32. Defferrard, M., Bresson, X., Vandergheynst, P.: Convolutional neural networks on graphs with fast localized spectral filtering. arXiv preprint arXiv:1606.09375 (2016)

33. Zhao, Z., Chen, W., Wu, X., Chen, P.C.Y., Liu, J.: LSTM network: a deep learning approach for short-term traffic forecast. IET Intel. Trans. Syst. **11**(2), 68–75 (2017)

Chapter 10
Time Series Classification Based on Complex Network

Kunfeng Qiu, Jinchao Zhou, Hui Cui, Zhuangzhi Chen, Shilian Zheng, and Qi Xuan

Abstract Time series classification plays an important role in various tasks such as electroencephalogram (EEG) classification, electrocardiogram (ECG) classification, human activity recognition and radio signal modulation identification. At present, some promising methods have been proposed to transform time series classification into graph classification by mapping time series to graphs. However, these transformation methods with fixed mapping rules may lack of flexibility, leading to the loss of information, so as to decrease the classification accuracy. In this chapter, we introduce circular limited penetrable visibility graph (CLPVG), a new method of mapping time series to graphs. Furthermore, in order to map time series to graphs more flexibly through deep learning, we also introduce an automatic visibility graph (AVG) based on graph neural network (GNN), a framework which can transform time series into graphs and realize classification end-to-end. Finally, we carry out experiments on some common datasets to prove the effectiveness of our methods.

10.1 Introduction

Time series are ubiquitous in almost all tasks requiring human cognitive processes [1], as a result, the classification of time series is a significant task in data mining [2]. For example, in the medical and health field, the EEG signals of epilepsy patients and autistic children are typical time series. By classifying the EEG signals of patients in different states, we can judge whether the patients are in a transitional state (from the normal moment to the onset moment), so as to take

K. Qiu · J. Zhou · H. Cui · Z. Chen · Q. Xuan (✉)
Institute of Cyberspace Security, Zhejiang University of Technology, Hangzhou, China

College of Information Engineering, Zhejiang University of Technology, Hangzhou, China
e-mail: xuanqi@zjut.edu.cn

S. Zheng
Science and Technology on Communication Information Security Control Laboratory, Jiaxing, China

corresponding measures in time to keep patients away from danger [3–7]. The ECG signals of patients with heart disease are also time series. By real-time monitoring and classification of the ECG signals of patients, emergency measures can be taken in time to protect patients before they have a heart attack. What's more, in radio signal modulation recognition, the received modulated signals are time series, and it is important to classify the modulation mode of the signals, which enables us to demodulate the modulated signals to obtain meaningful original signals from which some information can be obtained.

The traditional methods for time series classification can be divided into two processes, i.e., feature extraction and recognition. For example, in order to classify EEG signals and radio modulation signals, Soliman and Hsue [8] and Subasias [9] used Fourier transform [10] and wavelet transform [11] etc. to preprocess these signals, and then extract the high-order cycle spectrum, high-order cumulant, cyclo-stationary characteristics, power spectrum [12] and other characteristics of signals. Based on these features, traditional classification methods in machine learning, such as decision trees [13], random forest [14] and Support Vector Machine (SVM) [15], can be used to classify the time series. In general, the above mentioned process of analyzing the original signals requires much manpower and material resources [9], while the final classification accuracy is relatively low.

Recently, some advanced deep learning algorithms are also used to classify time series. For example, Hochreiter and Schmidhuber [16] proposed the Long Short-Term Memory (LSTM), a framework based on Recurrent Neural Network (RNN), which can classify time series with satisfactory accuracy. Based on Convolutional Neural Network (CNN), Wang et al. [17] proposed a time series classification model called Fully Convolutional Networks (FCN), while O'Shea et al. [18] used a residual neural network model for radio signal modulation recognition. Generally, the methods based on RNN and CNN give full play to the capability of deep learning models and can greatly improve the classification performance. However, RNN-based methods generally have high time complexity and require a lot of computation [19–21], while CNN-based methods ignore the temporal information in time series, which may lead to lower classification accuracy.

What's more, there are some methods based on fixed mapping rules to classify time series by transforming them into graphs. Lacasa et al. [22] first put forward the concept of Visibility Graph (VG), a novel idea of mapping time series to graphs. Since then, there have been some other mapping methods proposed based on VG, such as Horizontal Visibility Graph (HVG) [23] and Limited Penetrable Visibility Graph (LPVG) [24], etc. After converting time series into graphs, it is convenient for us to classify them with the help of graph classification methods. However, the above mapping methods are rigid, which can only convert each signal sample into a single graph. The original time series may not be well represented by the graphs derived by the fixed transformation rules.

In order to solve the above problem, we propose a more flexible transformation method based on LPVG, called circular limited penetrable visibility graph (CLPVG), which is able to map time series into suitable graphs that can better represent the corresponding signal samples. Furthermore, we propose an automatic visibility graph (AVG) based on graph neural network (GNN) to realize end-to-end time series classification. This method can generate and update graphs from time series by means of convolution operation and particular permutation, and can retain as much important information as possible from the original time series, thus further improving the classification accuracy. The corresponding experiments in this chapter prove that our two methods can effectively learn the graphs to represent time series, which are more flexible than the existing methods based on complex networks, leading to higher classification accuracy. And in theory, AVG can be more flexible and suitable to represent signals with graphs, since the mapping process can be adjusted automatically when it is integrated into the deep learning model.

The rest of this chapter is organized as follows. In Sect. 10.2, we briefly describe the related work on time series classification and graph classification, and then introduce the LVPG, a typical mapping method for converting time series to graphs. In Sect. 10.3, we introduce our two methods CLPVG and AVG in detail. In Sect. 10.4, we give the experiments to validate the effectiveness of our two methods. Finally, we give the conclusion in Sect. 10.5.

10.2 Related Work

10.2.1 Time Series Classification

Time series classification is increasingly important in data mining and other fields, which has been applied to automatic speech recognition (ASR) [25], electrocardiogram (ECG) classification and chemical engineering [2, 26]. Suppose a time series sample Y of length n:

$$Y = \{y_1, y_2, \cdots, y_k, \cdots, y_n\} \tag{10.1}$$

where y_k represents the value corresponding to time point k. Let (k, y_k) represent a signal sampling point (k, y_k). And the purpose of time series classification is to predict the category labels of new time series by the model established on the training data with class labels.

Before the rise of deep learning, traditional time series classification methods were mainly based on the handcraft features together with the typical machine learning methods. For example, Nanopoulos et al. [27] extracted statistical features such as the mean value and the variance of time series, and then completed classification with Multilayer Perceptron (MLP). Deng et al. [28] proposed Time Series Forest (TSF) which combines entropy gain and distance metric. What's more, there are also some distance-based classification methods to predict the category of a test sample. These methods first use distance function to calculate the similarity between the current sample and all training samples, and then the label of the training sample with the smallest distance value is considered to be the category of the current test sample. The distance function used in these methods generally includes Dynamic Time Warping (DTW) [29], Edit Distance (ED) [30] and Longest Common Sub-sequence (LCS) [31], etc. In general, the features extracted by these classification methods are representative and can represent the original signals well. However, it is also obvious that these methods require a solid level of expertise.

With the rapid development of deep learning, neural networks have been widely used in time series classification and achieved gratifying results. Wang et al. [32] proposed Fully Convolutional Neural Networks (FCN), a CNN-based classification model, simple but effective, which can classify time series without complex feature extraction process. Karim et al. [33] proposed LSTM-FCN and ALSTM-FCN, two end-to-end models for time series classification, which combine CNN and RNN. Compared with the traditional methods, the classification algorithms based on deep learning make use of big data to learn features and can better capture rich internal information of time series.

10.2.2 Mapping Methods

As is known to all, CNN model is originally designed to deal with images, but it is surprising that its performance in time series classification is also excellent. Inspired by this, some researchers tried to map time series to graphs, and then use the technologies in complex network to analyze time series. With the advent of Visibility Graph (VG) [22], a method that maps time series to graphs, more and more similar mapping methods emerge, among which Limited Penetrable Visibility Graph (LPVG) [24] is a well-known one that can be combined with commonly used classifiers to classify time series with good performance. At first, LPVG changes the time series Y to the form of a vertical line diagram. In order to ensure that the height of each vertical line is positive, the original time series Y is processed as follows:

$$Y' = \{y_1 + a, y_2 + a, \cdots, y_k + a, \cdots, y_n + a\}, a > |\min(Y)|, \quad (10.2)$$

where the minimum value in time series Y is denoted as $min(Y)$, and a is a number greater than the absolute value of $min(Y)$. Take the n time points as the abscissa, and the processed signal values corresponding to the time points as the ordinate to construct the vertical line diagram. And then, according to the obtained vertical lines, LPVG judges whether every two points Y_i and Y_j can constitute an edge in turn, where $1 \leq i \leq j \leq n$. The specific operation is to connect the tops of the vertical lines corresponding to the sampling points Y_i and Y_j to obtain a line L_{ij}. The number of intersections between the line L_{ij} and the vertical lines corresponding to the sampling points Y_m is denoted as L, where $m = i + 1, \ldots, j - 1$. Given a hyperparameter b, if L is less than or equal to b, then it is considered that the sampling points Y_i and Y_j can form an edge in the graph. According to the above rule, the time series Y with length n can finally be mapped into a graph G with n nodes. Take a time series of length 5 as an example, and set the hyperparameter

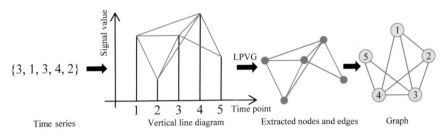

Fig. 10.1 The vertical line diagram of time series with length 5 and the graph obtained by LPVG

b to 1, the constructed vertical line diagram and the mapped graph are shown in Fig. 10.1.

10.2.3 Graph Classification

After mapping time series to graphs, the problem of time series classification is naturally transformed into graph classification. At present, the advanced methods for graph classification mainly include graph embedding and graph neural network [34–37]. In graph classification, graph embedding is used to represent a graph with a low-dimensional vector, and then some existing machine learning classification algorithms are adopted to realize the classification. On the one hand, we can artificially extract some attributes of a graph to form a low-dimensional vector, such as calculating the number of nodes, the number of edges and clustering coefficient of the network and so on [38]. On the other hand, some unsupervised algorithms can also be used to automatically extract feature vectors of networks, such as Graph2vec [39], which is a typical graph embedding method. Graph2vec is the first unsupervised embedding method for the entire network. Note that graph2vec uses a model similar to doc2vec [40] which is based on extended text and embedding technology and shows great advantages in Natural Language Processing (NLP). Similarly, graph2vec establishes a relationship between the network and the rooted subgraph. Graph2vec first extracts the rooted subgraph and provides the corresponding label into the vocabulary, and then trains the Skip-Gram model to obtain the representation of the entire network. This method performs well in graph classification tasks, outperforming most manually feature extraction methods.

In addition, with the development of deep learning, there emerges many end-to-end GNN models that can accomplish graph classification, such as Graphsage [41] and Diffpool [42]. The GNN can embed the nodes based on their local neighborhood information, i.e., aggregating the information of each node and its surrounding nodes. In particular, the process of aggregating information can be optimized through neural networks. In our AVG, we use Diffpool as the basic graph classification model. Diffpool coarsens graph based on the output of operation layer in

GNN, and then maps the nodes into some clusters which are inputs of the next operation layer in GNN. As a differentiable graph-level pooling module, Diffpool is able to generate hierarchical representation of graphs to classify them [42].

10.3 Methods

In this section, we mainly introduce our two time series classification methods based on graphs, CLPVG and AVG, and discuss the differences between them and LPVG.

10.3.1 Circular Limited Penetrable Visibility Graph

There are some methods to analyze time series by converting them into graphs, such as Visibility Graph (VG) [22], Limited Penetrable Visibility Graph (LPVG) [24] and so on. Experiments [22, 24] have proved that these methods of building graphs from time series can retain and extract some basic features of time series. However, the graphs obtained by these algorithms are relatively simple, and it is almost impossible for them to construct graphs containing more effective information from the time series according to different tasks or user requirements. Therefore, based on the existing LPVG method, we propose a new method, circular limited penetrable visibility graph (CLPVG), to convert time series to graphs. This method is more flexible and can extract more effective information from time series, so as to achieve better classification results.

In this section, we mainly introduce the specific process of time series classification by means of CLPVG. In general, it is an adjustable nonlinear graph construction algorithm, and it is also a visibility graph construction algorithm that uses arcs instead of straight lines in LPVG when constructing graphs, with the aim to completing time series classification task better. The overall process of time series classification through method CLPVG is shown in Fig. 10.2. In this process, time series are first converted into graphs by CLPVG, then the first-order subgraph networks [43] are extracted. After that, all graphs are processed by Graph2vec [39] to obtain corresponding feature vectors, which are then concatenated and classified by a classifier in the field of machine learning.

10.3.1.1 Circle System Equation

In mathematics, a set of circles meeting certain conditions are called a circle system, and the equations describing a circle system are called circle system equations. As shown in Fig. 10.3, given any two different data points (t_1, y_1) and (t_2, y_2), we can

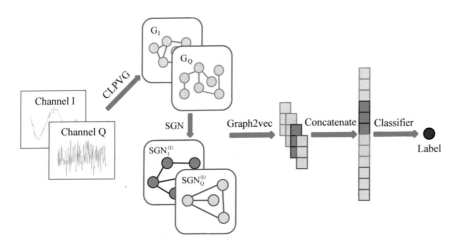

Fig. 10.2 The overall process of time series classification through CLPVG

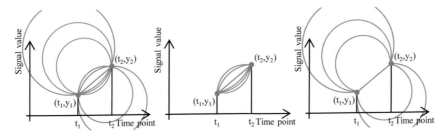

Fig. 10.3 Schematic diagram of circle system. When LPVG is used to transform time series into graphs, the blue line is used to connect every two data points so as to judge whether the two points can form edge. When using CLPVG, we use one of the orange arcs instead of the blue line to build graph. As for the orange arcs, we can choose either the shorter arc line part or the longer arc line part, as shown in the center and the right picture, respectively

get many different circles, and the circle system equation can be expressed as:

$$f(t, y) = (t - t_i)(t - t_j) + (y - y_i)(y - y_j)$$
$$+ a[(t - t_i)(y_j - y_i) - (y - y_i)(t_j - t_i)] = 0,$$
(10.3)

where a is a hyperparameter that controls the size of the circle. Obviously, the two points (t_1, y_1) and (t_2, y_2) divide the circle formed by them into two parts, one is the longer arc line part, and the other is the shorter part. In the subsequent work of constructing the network, we choose the shorter arc line to replace the line of method LVPG.

10.3.1.2 Graph Construction through CLPVG

Similar to method LPVG, CLPVG only changes straight line segments to shorter arcs, and keeps the other rules the same. Finally, the time series can be mapped into graphs. Take the time series shown in Fig. 10.1 as an example, assuming the hyperparameter $b = 1$, the graph construction through CLPVG is shown in Fig. 10.4.

The pseudocode of mapping time series Y to graph $G = \langle V, E \rangle$ by CLPVG is shown in Algorithms 1. We treat all time points in Y as nodes of graph G at first. And then, we determine whether every two nodes can form an edge in G.

Algorithm 1: Map time series to graph through CLPVG

Input: The time series Y as shown in Eq. (10.1), length n of Y, hyperparameter a shown in Eq. (10.3), hyperparameter b mentioned in Sect. 10.2.2.

Output: Graph $G = \langle V, E \rangle$ formed by CLPVG.

1 Set the value of the hyperparameter a and b

2 for $ln=1$ to $n-1$ **do**

3   Append node ln to V

4 $add = 0$

5   **for** $rn=ln+1$ to n **do**

6   **for** $mn=ln$ to rn **do**

7   Compute the y'_{mn} when $t = mn$ via Eq. (10.3)

8 $add = add + 1$ if $y_{mn} \geq y'_{mn}$

9   Append edge (ln, rn) to E if $add \leq b$

10 Append node n to V

11 return $G = \langle V, E \rangle$

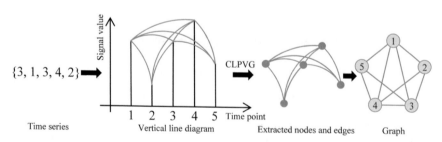

$\{3, 1, 3, 4, 2\}$ ⟹

CLPVG ⟹

Time series   Vertical line diagram   Extracted nodes and edges   Graph

Fig. 10.4 The graph obtained by CLPVG. Here we use the same time series in Fig. 10.1 for comparison

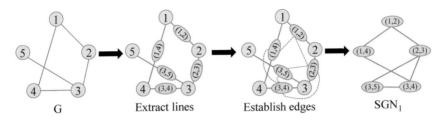

Fig. 10.5 The process of build SGN$^{(1)}$ from the original graph G

10.3.1.3   Subgraph Network

After mapping time series into graphs, we can use certain technologies in the network field to improve the final classification accuracy. Here we adopt subgraph network (SGN) [43] to expand feature space and further enhance the graph classification algorithms, which is carefully introduced in Chap. 3. Given a graph $G = \langle V, E \rangle$, where V and E respectively represent the sets of nodes and edges of graph G, we extract the first-order subgraph networks SGN$^{(1)}$ from G, as shown in Fig. 10.5. At first, we map the edges in the original graph G into the nodes in SGN$^{(1)}$. Then, if the two edges in G share the same terminal node, then the two corresponding nodes in SGN$^{(1)}$ mapped from these two edges are connected. Certainly, we can use a similar way to get SGN of higher orders iteratively. In general, SGN can dig out some deeper features in the original graph, which helps to improve the graph classification accuracy. However, extracting subgraph networks of higher order could significantly increase the consumption of resources and time, but may not do much to improve the accuracy, so we only use SGN$^{(1)}$ in the experiments.

We further use Graph2vec [39] to extract features of the generated graphs, as well as the corresponding first-order SGNs, and then use random forest [14] to realize the classification.

10.3.2   Automatic Visibility Graph based on GNN

In this section, we mainly introduce our automatic visibility graph (AVG) based on GNN, an end-to-end deep learning framework. Different from the above-mentioned methods of mapping time series into graphs according to fixed rules, AVG can obtain graphs through self-learning and realize time series classification by classifying the obtained graphs in a deep learning framework.

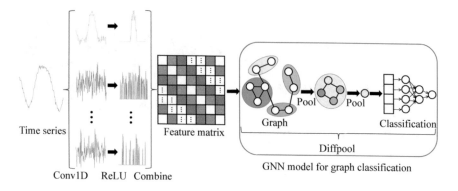

Fig. 10.6 The overall framework of AVG. Here, we choose the Diffpool model as our classifier

10.3.2.1 The Overall Framework

The idea of this method is to process the time series by multiple one-dimensional convolution layers with different convolution kernel sizes to obtain the processed feature sequences, and then to sort them into a feature matrix according to certain rules which can represent a graph. Finally, the graph classification model in the field of GNN is used to classify the obtained graphs. In this chapter, we choose the typical GNN model Diffpool [42] to classify the graphs. Note that other GNN models for graph classification, such as GraphSage [41], could also be used in the future. In particular, the processing of time series by one-dimensional convolution layers can be trained together with the graph classification model. This transformation method of mapping time series into graphs thus is very flexible and not limited to fixed rules, so it has universal applicability to time series of different fields. The overall framework of AVG is shown in Fig. 10.6.

10.3.2.2 Feature Extraction

Firstly, suppose that each sample in a given time series dataset can be expressed as:

$$t_{1 \times N} = [t_1, t_2, \ldots, t_N], \tag{10.4}$$

where N represents the number of time points in the time series, t_N represents the value corresponding to the N-th time point, and $t_{1 \times N}$ represents the time series sample of length N. And then, the original time series are processed by one-dimensional convolution layers with multiple convolution kernels of different sizes to obtain the corresponding processed feature sequences. The specific convolution process is as follows:

$$\phi_m = Conv1D_m(t_{1 \times N}) = [x_1^m, x_2^m, \ldots, x_{N+1-m}^m], 2 \leq m \leq k, \tag{10.5}$$

where k is used to control the number of one-dimensional convolutional layers, $Conv1D_m(\cdot)$ denotes the one-dimensional convolution layer with convolution kernel size m and step size 1. ϕ_m represents the feature sequence of length $N + 1 - m$ obtained after processing the original time series with one-dimension convolutional layer $Conv1D_m(\cdot)$, where the i-th feature is expressed as x_i^m. Then, all the obtained feature sequences are processed with ReLU activation function to obtain the processed feature sequences whose feature elements are all non-negative numbers.

$$\varphi_m = ReLU(\phi_m) = [y_1^m, y_2^m, \cdots, y_{N+1-m}^m], 2 \leq m \leq k, \tag{10.6}$$

10.3.2.3 Feature Matrix of Graph

After extracting the feature sequences, we process these features to obtain the feature matrices of graphs. The obtained non-negative feature sequences are sorted according to a certain rule to obtain the feature matrix M that can represent the network of size $N \times N$. The sorting rule is to place the non-negative feature value y_i^m in the i-th row and the $(i + m - 1)$-th column, as well as the $(i + m - 1)$-th row and the i-th column of M, and the other positions of matrix M are 0. The resulting feature matrix M is as follows:

$$M = \begin{bmatrix} 0 & y_1^2 & \cdots & y_1^k & 0 & 0 \\ y_1^2 & 0 & y_2^2 & \cdots & \ddots & 0 \\ \vdots & y_2^2 & 0 & y_3^2 & \ddots & y_{N+1-k}^k \\ y_1^k & \vdots & y_3^2 & 0 & \ddots & \vdots \\ 0 & \ddots & \ddots & \ddots & \ddots & y_{N-1}^2 \\ 0 & 0 & y_{N+1-k}^k & \cdots & y_{N-1}^2 & 0 \end{bmatrix}, \tag{10.7}$$

Especially, because there is no self-connection, the diagonal element values in the feature matrix of the graph are all equal to 0. It is obvious that the feature sequence obtained by processing the original time series with the one-dimensional convolution layer with kernel length 1 will be placed on the diagonal of the matrix M, therefore, the convolution kernel length of the one-dimensional convolution layer needs to be greater than 1. And this shows that the graphs constructed by method AVG are weighted and un-directed graphs. We summarize the process of mapping time series N to graph $G = \langle V, E \rangle$ in Algorithm 2. At first, we use some one-dimensional convolutional layers with different kernel sizes and the activation function ReLU to process the original time series. After obtaining the corresponding feature vectors, we combine them into a feature matrix. Finally, it is easy for us to transform the symmetric feature matrix into a graph.

Algorithm 2: Map time series to graph by AVG

Input: The time series $t_{1\times N}$ as shown in Eq. (10.4), length N of $t_{1\times N}$, hyperparameter k shown in Eq. (10.5).

Output: Graph $G = \langle V, E \rangle$ formed in AVG.

1 Create a zero matrix M with shape $N \times N$
2 **for** $m=2$ to k **do**
3   Define an one-dimensional convolutional layer $Conv1D_m(\cdot)$ with convolution kernel
4   size m and step size 1
5   Compute the feature sequence ϕ_m via Eq. (10.5)
6   Compute the non-negative feature sequence $\varphi_m = [y_1^m, y_2^m, \cdots, y_{N+1-m}^m]$ via
    Eq. (10.6)
7   **for** $i=1$ to $N+1$-m **do**
8          $M(i, i+m-1) = M(i, i+m-1) = y_i^m$ where $M(r, c)$ represents the element in
        the r-th row and the c-th column of M

9 Construct the weighted undirected graph according to M
10 **return** $G =< V, E >$

10.3.2.4 Classification of Graphs

After the weighted un-directed graphs are obtained from the time series, the neural network model used for graph classification, such as GraphSage [41] and Diffpool [42], can be used to classify the obtained graphs, so as to achieve the purpose of time series classification. In particular, the one-dimensional convolution operation that maps the time series into graphs can be trained with the used GNN model, so as to get the most suitable graphs.

10.3.3 Comparison with LPVG

The only difference between CLPVG and LPVG is that we use arc rather than line to build graphs, as shown in Figs. 10.4 and 10.1. Arcs are more flexible than lines, since different arc bending degrees can be set up to construct different graphs, i.e., the graphs that mostly represent the original time series can be obtained by choosing the appropriate arc bending degree.

In addition, AVG does not map time series to graphs according to fixed rules, but rather designs a mapping by a self-learning mechanism, which is even more flexible than CLPVG. From the point of mapping rules for constructing graphs by LPVG, when determining whether there is an edge between two nodes $Y_i = (i, Y_i)$ and $Y_j = (j, Y_j)$, it is necessary to construct a line L_{ij} passing through these two points in the vertical line graph. The line L_{ij} is represented as:

$$f(t) = \frac{y_j - y_i}{j - i} \times (t - i) + y_i, \tag{10.8}$$

then the value y_m of each time point m between i and j and the value $f(m)$ of the linear function L_{ij} at time m are compared to determine whether $Y_i = (i, y_i)$ and $Y_j = (j, y_j)$ can be connected in the graph. To some extent, this method ignores the specific time series values, which results in the loss of some hidden information in the original time series when the time series are mapped into graphs, and finally affects the classification accuracy. Unlike LPVG, AVG uses one-dimensional convolution layer to process the relationship between every two nodes, and calculate the correlation degree of these two nodes and the nodes between them. It makes full use of the specific time series value corresponding to each time point and can retain more information of the original time series. In addition, the graphs constructed by LPVG are unweighted, while our AVG can construct the weighted networks. Therefore, more features in the original time series can be extracted and the classification accuracy thus can be improved.

10.4   Experiments

In this section, we mainly introduce the specific implementation processes and the results of CLPVG and AVG, and compare with those obtained by LPVG on some public datasets, to validate the effectiveness of our methods.

10.4.1   Datasets

The datasets used in the experiments include time series in wireless radio modulation (RML2016.10a [44]) and medical field (EEG [45]), and three typical UCR time series (Adiac, ElectricDevices and Herring) [46].

RML2016.10a is a high-quality radio signal simulation dataset generated based on the GNU Radio environment. It contains 11 modulation types. Each signal of each modulation type contains 20 kinds of signal-to-noise ratios (SNRs), each SNR has 1000 signal samples, each sample has two channels I and Q, and the signal length of each channel is 128. We divide the SNR data of each modulation type by a ratio of 4:1 to get the training set and the test set.

As for the EEG dataset, it is composed of 5 sub-datasets A, B, C, D, and E, among which subsets A and B are the EEG signals of healthy people with eyes open and closed respectively. Subsets C, D, and E are collected in patients with epilepsy. Especially, we divide the EEG dataset into EEG^1 and EEG^2, where EEG^1 regards the subset E as a single category, the remaining four subsets as another category, and EEG^2 regards each subset as a separate category.

Moreover, we also choose time-series datasets from UCR database [46] to testify our methods as well as LPVG. Specifically, we use the Adiac, ElectricDevices and Herring datasets here.

The basic statistics of the above datasets are presented in Table 10.1.

Table 10.1 The basic statistics of the datasets adopted in the experiments

Dataset	# Training samples	# Testing samples	# Classes	Length
RML2016.10a	176,000	44,000	11	128
EEG[1]	450	50	2	4097
EEG[2]	450	50	5	4097
Adiac	390	391	37	176
ElectricDevices	8926	7711	7	96
Herring	64	64	2	512

10.4.2 The Experimental Settings

When verifying the method CLPVG, since each sample of dataset RML2016.10a has two channels, we process it as shown in Fig. 10.2. At first, we process the I channel and Q channel data of each signal to obtain the graphs G_I and G_Q respectively. Then, we extract the first-order subgraph networks $SGN_I^{(1)}$ and $SGN_Q^{(1)}$ of the graphs G_I and G_Q, respectively, so as to represent a signal sample with four graphs G_I, G_Q, $SGN_I^{(1)}$ and $SGN_Q^{(1)}$. Then, we use the graph embedding method Graph2vec to extract four feature vectors of length L for these graphs. We further combine them to get a single vector of length $4L$ to represent a signal sample. Finally, we use the random forest [14] to process the feature matrix composed of all feature vectors extracted form all signals to complete the classification. As for the other datasets, since they are all univariate time series, we only need to convert each sample into a graph, get the corresponding $SGN^{(1)}$, and then use two, instead of four, graphs to represent each signal sample. The other settings are also the same.

As for the method AVG, it is easy for us to process the univariate time series EEG, Adiac, ElectricDevices and Herring according to the process shown in Fig. 10.6. However, we can not directly process the multivariate time series dataset RML2016.10a, because Diffpool requires that the number of graphs for each input sample is 1. As a result, we delete the last fully connected layer of the Diffpool model, and then use this incomplete Diffpool model to process the graphs that obtained by the I channel and Q channel separately and get the corresponding two

Table 10.2 The classification accuracy of LPVG, CLPVG and AVG on different datasets. The bold values are the best results

Dataset	LPVG	LPVG+SGN[(1)]	CLPVG	CLPVG+SGN[(1)]	AVG
RML2016.10a	47.68%	48.48%	49.40%	50.07%	**55.36%**
EEG[1]	97.75%	97.60%	97.80%	97.80%	**100%**
EEG[2]	64.82%	72.62%	67.40%	73.42%	**76.00%**
Adiac	42.20%	44.05%	69.05%	**72.12%**	64.45%
ElectricDevices	64.42%	64.80%	65.82%	66.54%	**71.49%**
Herring	70.31%	64.06%	70.31%	65.63%	**71.88%**

feature vectors. After concatenating these two vectors into a single one, we then use a fully connected layer for classification.

Since our methods CLPVG and AVG are both inspired by LPVG, here we simply use LPVG as the baseline for comparison.

10.4.3   The Experimental Results

The results of using LPVG, CLPVG and AVG to classify the time series are shown in Table 10.2. In order to see the impact of SGN on the methods LPVG and CLPVG, we divide the two methods into the cases of using SGN and not using SGN. In particular, we give the accuracy of different methods under different signal-to-noise ratios of the RML2016.10a dataset in Table 10.3. The purpose of proposing CLPVG is to map time series to graphs more flexibly than LPVG to cope with specific

Table 10.3 The accuracy under different signal-to-noise ratios of the RML2016.10a dataset. The bold values are the best results

SNR	LPVG	LPVG+SGN[1]	CLPVG	CLPVG+SGN[1]	AVG
18 dB	78.77%	80.77%	79.77%	79.77%	**83.82%**
16 dB	77.05%	79.50%	78.50%	79.27%	**83.86%**
14 dB	74.32%	78.86%	78.09%	79.77%	**82.59%**
12 dB	75.73%	80.41%	78.09%	77.77%	**83.09%**
10 dB	75.45%	79.36%	77.18%	78.05%	**85.23%**
8 dB	74.18%	77.05%	74.68%	76.09%	**82.45%**
6 dB	69.27%	74.82%	73.68%	72.73%	**81.23%**
4 dB	66.68%	66.55%	67.59%	70.23%	**82.82%**
2 dB	57.09%	55.86%	59.36%	62.36%	**79.82%**
0 dB	52.77%	46.55%	53.18%	56.45%	**76.00%**
−2 dB	43.68%	42.32%	42.55%	48.55%	**66.64%**
−4 dB	39.95%	41.55%	37.14%	42.09%	**55.50%**
−6 dB	33.64%	38.45%	37.09%	36.95%	**45.05%**
−8 dB	28.41%	29.05%	30.91%	28.55%	**32.68%**
−10 dB	21.64%	19.45%	24.09%	23.14%	**24.23%**
−12 dB	18.64%	17.59%	**21.73%**	19.59%	16.68%
−14 dB	16.86%	15.91%	**20.32%**	17.64%	14.91%
−16 dB	16.45%	14.86%	**18.05%**	17.55%	10.00%
−18 dB	16.77%	15.32%	18.14%	**18.14%**	10.45%
−20 dB	16.27%	15.36%	**17.77%**	16.68%	10.14%
Total	47.68%	48.48%	49.40%	50.07%	**55.36%**
≥0 dB	70.13%	71.97%	72.01%	73.25%	**82.09%**
>5 dB	74.97%	78.68%	77.14%	77.64%	**83.18%**
> −5 dB	65.41%	66.97%	66.65%	68.59%	**78.59%**

classification tasks, such as improving the classification accuracy of a specific signal-to-noise ratio or a specific range of signal-to-noise ratio signals. And the intention of designing AVG is to get a even better graph representation of the time series, leading to much better classification accuracy.

It can be seen from the Tables 10.2 and 10.3 that the classification accuracy obtained by CLPVG is indeed slightly higher than that of LPVG by comparing the results in the second and fourth columns. Similarly, by comparing the results in the third and the fifth columns, we can also get the same conclusion. Such results validate that, thanks to CLPVG's ability to control the bending degree of the arc, the graphs obtained by this mapping method can represent the original time series better than those obtained by LPVG. Moreover, in most cases, SGN has a positive effect on improving the classification accuracy, for both CLPVG and LPVG.

Besides, it is obvious that AVG significantly outperforms CLPVG and LPVG on all the considered datasets except for Adiac, no matter whether SGN is adopted or not. This result is quite reasonable since AVG based on GNN is much more flexible to establish graphs for time series. In other words, these graphs obtained by AVG are somewhat optimized to classify time series, and thus can better capture the hidden features of the original time series. More interestingly, it seems that CLPVG perform surprisingly better than AVG and LPVG for the signals of low SNR, as shown in Table 10.3, which suggest that CLPVG may be good candidate to process the time series containing a lot of noises to get robust results.

10.5   Conclusion

In this chapter, we focus on mapping time series to graphs, and propose two novel mapping methods, CLPVG and AVG, which are more flexible than the well-known LPVG and the graphs generated by these two methods thus can better capture the latent features of original time series. In particular, we replace the lines in LPVG by arcs in CLPVG through introducing a hyperparameter to control the arc bending degree, which makes CLPVG more flexible than LPVG to process different kinds of time series. For AVG, we further use GNN to generate graphs for time series automatically and finish time series classification end-to-end, achieving the state-of-the-art performance. Moreover, the SGN method introduced in Chap. 3 is also used here to expand the feature space, and enhance the performance of LPVG and CLPVG to certain extent. In the future, we will try to develop more methods to map time series to graphs, so as to facilitate the analysis of time series by utilizing advanced graph data mining algorithms.

References

1. Wong, M.D., Nandi, A.K.: Automatic digital modulation recognition using artificial neural network and genetic algorithm. Signal Process. **84**(2), 351–365 (2004)
2. Dobre, O.A., Abdi, A., Bar-Ness, Y., Su, W.: Survey of automatic modulation classification techniques: classical approaches and new trends. IET Commun. **1**(2), 137–156 (2007)
3. Ahmadlou, M., Adeli, H., Adeli, A.: Improved visibility graph fractality with application for the diagnosis of autism spectrum disorder. Physica A: Stat. Mech. Appl. **391**(20), 4720–4726 (2012)
4. Wang, J., Yang, C., Wang, R., Yu, H., Cao, Y., Liu, J.: Functional brain networks in alzheimer's disease: Eeg analysis based on limited penetrable visibility graph and phase space method. Physica A: Stat. Mech. Appl. **460**, 174–187 (2016)
5. Gao, Z.-K., Cai, Q., Yang, Y.-X., Dang, W.-D., Zhang, S.-S.: Multiscale limited penetrable horizontal visibility graph for analyzing nonlinear time series. Sci. Rep. **6**, 35622 (2016)
6. Pei, X., Wang, J., Deng, B., Wei, X., Yu, H.: WLPVG approach to the analysis of EEG-based functional brain network under manual acupuncture. Cogn. Neurodyn. **8**(5), 417–428 (2014)
7. Supriya, S., Siuly, S., Wang, H., Cao, J., Zhang, Y.: Weighted visibility graph with complex network features in the detection of epilepsy. IEEE Access **4**, 6554–6566 (2016)
8. Soliman, S.S., Hsue, S.-Z.: Signal classification using statistical moments. IEEE Trans. Commun. **40**(5), 908–916 (1992)
9. Subasi, A.: Eeg signal classification using wavelet feature extraction and a mixture of expert model. Expert Syst. Appl. **32**(4), 1084–1093 (2007)
10. Bracewell, R.N., Bracewell, R.N.: The Fourier Transform and its Applications, vol. 31999 (McGraw-Hill, New York, 1986)
11. Antonini, M., Barlaud, M., Mathieu, P., Daubechies, I.: Image coding using wavelet transform. IEEE Trans. Image Process. **1**(2), 205–220 (1992)
12. B. Mollow, Power spectrum of light scattered by two-level systems. Phys. Rev. **188**(5), 1969–1975 (1969)
13. Quinlan, J.R.: Induction of decision trees. Mach. Learn. **1**(1), 81–106 (1986)
14. Liaw, A., Wiener, M. et al.: Classification and regression by randomforest. R news **2**(3), 18–22 (2002)
15. Suykens, J.A., Vandewalle, J.: Least squares support vector machine classifiers. Neural. Process. Lett. **9**(3), 293–300 (1999)
16. Hochreiter, S., Schmidhuber, J.: Long short-term memory. Neural Comput. **9**(8), 1735–1780 (1997)
17. Wang, Z., Yan, W., Oates, T.: Time series classification from scratch with deep neural networks: A strong baseline. In: Proceedings of the 2017 International Joint Conference on Neural Networks (IJCNN), pp. 1578–1585 (IEEE, New York, 2017)
18. O'Shea, T.J., Corgan, J., Clancy, T.C.: Convolutional radio modulation recognition networks. In: International Conference on Engineering Applications of Neural Networks, pp. 213–226 (Springer, Berlin, 2016)
19. Zaremba, W., Sutskever, I., Vinyals, O.: Recurrent neural network regularization. arXiv preprint arXiv:1409.2329 (2014)
20. Hüsken, M., Stagge, P.: Recurrent neural networks for time series classification. Neurocomputing **50**, 223–235 (2003)
21. Malhotra, P., TV, V., Vig, L., Agarwal, P., Shroff, G.: Timenet: Pre-trained deep recurrent neural network for time series classification. arXiv preprint arXiv:1706.08838 (2017)
22. Lacasa, L., Luque, B., Ballesteros, F., Luque, J., Nuno, J.C.: From time series to complex networks: the visibility graph. Proc. Natl. Acad. Sci. **105**(13), 4972–4975 (2008)
23. Xie, W.-J., Han, R.-Q., Zhou, W.-X.: Tetradic motif profiles of horizontal visibility graphs. Commun. Nonlinear Sci. Numer. Simul. **72**, 544–551 (2019)
24. Cai, L., Wang, J., Cao, Y., Deng, B., Yang, C.: LPVG analysis of the EEG activity in alzheimer's disease patients, in *Proceedings of the 2016 12th World Congress on Intelligent Control and Automation (WCICA)* (2016)

25. Lee, K.-F.: *Automatic Speech Recognition: The Development of the SPHINX System*, vol. 62 (Springer, Berlin, 1988)
26. Fawaz, H.I., Forestier, G., Weber, J., Idoumghar, L., Muller, P.-A.: Deep learning for time series classification: a review. Data Min. Knowl. Disc. **33**(4), 917–963 (2019)
27. Nanopoulos, A., Alcock, R., Manolopoulos, Y.: Feature-based classification of time-series data. Int. J. Comput. Res. **10**(3), 49–61 (2001)
28. Deng, H., Runger, G., Tuv, E., Vladimir, M.: A time series forest for classification and feature extraction. Inf. Sci. **239**, 142–153 (2013)
29. Berndt, D.J., Clifford, J.: Using dynamic time warping to find patterns in time series. In: KDD Workshop, vol. 10, pp. 359–370, Seattle, WA, USA (1994)
30. Ristad, E.S., Yianilos, P.N.: Learning string-edit distance. IEEE Trans. Pattern Anal. Mach. Intell. **20**(5), 522–532 (1998)
31. Hunt, J.W., Szymanski, T.G.: A fast algorithm for computing longest common subsequences. Commun. ACM **20**(5), 350–353 (1977)
32. Wang, Z., Yan, W., Oates, T.: Time series classification from scratch with deep neural networks: A strong baseline. In: Proceedings of the 2017 International Joint Conference on Neural Networks (IJCNN), pp. 1578–1585 (2017)
33. Karim, F., Majumdar, S., Darabi, H., Chen, S.: LSTM fully convolutional networks for time series classification. IEEE Access **6**, 1662–1669 (2017)
34. Zhang, M., Cui, Z., Neumann, M., Chen, Y.: An end-to-end deep learning architecture for graph classification. In: Proceedings of the Thirty-Second AAAI Conference on Artificial Intelligence (2018)
35. Lee, J.B., Rossi, R., Kong, X.: Graph classification using structural attention. In: Proceedings of the 24th ACM SIGKDD International Conference on Knowledge Discovery and Data Mining, pp. 1666–1674 (2018)
36. Riesen, K., Bunke, H.: Graph classification based on vector space embedding. Int. J. Pattern Recognit. Artif. Intell. **23**(06), 1053–1081 (2009)
37. Kudo, T., Maeda, E., Matsumoto, Y.: An application of boosting to graph classification. In: Advances in Neural Information Processing Systems, pp. 729–736 (2005)
38. Costa, L.d.F., Rodrigues, F.A., Travieso, G., Boas, P.R.V.: Characterization of Complex Networks: A Survey of Measurements (2005)
39. Narayanan, A., Chandramohan, M., Venkatesan, R., Chen, L., Liu, Y., Jaiswal, S.: graph2vec: learning distributed representations of graphs. arXiv preprint arXiv:1707.05005 (2017)
40. Lau, J.H., Baldwin, T.: An empirical evaluation of doc2vec with practical insights into document embedding generation. arXiv preprint arXiv:1607.05368 (2016)
41. Hamilton, W., Ying, Z., Leskovec, J.: Inductive representation learning on large graphs. In: Advances in Neural Information Processing Systems, pp. 1024–1034 (2017)
42. Ying, Z., You, J., Morris, C., Ren, X., Hamilton, W., Leskovec, J.: Hierarchical graph representation learning with differentiable pooling. In: Advances in Neural Information Processing Systems, pp. 4800–4810 (2018)
43. Xuan, Q., Wang, J., Zhao, M., Yuan, J., Fu, C., Ruan, Z., Chen, G.: Subgraph networks with application to structural feature space expansion. IEEE Trans. Knowl. Data Eng. **33**(6), 2776–2789 (2021)
44. O'Shea, T.J., Corgan, J., Clancy, T.C.: Convolutional radio modulation recognition networks. In: International Conference on Engineering Applications of Neural Networks, pp. 213–226 (Springer, Berlin, 2016)
45. Andrzejak, R.G., Lehnertz, K., Mormann, F., Rieke, C., David, P., Elger, C.E.: Indications of nonlinear deterministic and finite-dimensional structures in time series of brain electrical activity: Dependence on recording region and brain state. Phys. Rev. E **64**(6), 061907 (2001)
46. Dau, H.A., Bagnall, A., Kamgar, K., Yeh, C.-C.M., Zhu, Y., Gharghabi, S., Ratanamahatana, C.A., Keogh, E.: The UCR time series archive. IEEE/CAA J. Automat. Sin. **6**(6), 1293–1305 (2019)

Chapter 11
Exploring the Controlled Experiment by Social Bots

Yong Min, Yuying Zhou, Tingjun Jiang, and Ye Wu

Abstract With the continuous development of digital media, people receive more and more information on the Internet. The emergence and development of social bots, on the one hand, serve human beings, on the other hand, cause the pollution of the network environment. Therefore, more and more researches begin to focus on social bots. In this chapter, we first introduce the concept of social bots, including the definition, application and influence. Then we introduce the needed technologies to deploy social bots on social networks and summarize four methods to detect social bots. We mainly analyze the feasibility of social bot as a controlled experiment to study social network, and review some existing research results.

11.1 Introduction

Nowadays, with the development of digital media, social bots are widely used in various networks. With the gradual maturity of artificial intelligence, social bots become more and more "smart" to assist "masters" to complete various tasks. The original purpose of social bots is to serve human beings and society. But there are also malicious social bots that disrupt the stability of networks for a particular purpose. Because of the increasing proportion of digital media, most of people's information comes from the online network, the influence of these malicious bots is

Y. Min (✉)
Institute of Cyberspace Security, Zhejiang University of Technology, Hangzhou, China

College of Information Engineering, Zhejiang University of Technology, Hangzhou, China
e-mail: myong@zjut.edu.cn

Y. Zhou · T. Jiang
College of Computer Science and Technology, Zhejiang University of Technology, Hangzhou, China

Y. Wu
Beijing Normal University, Zhuhai, China

becoming more and more powerful. However, we can not ignore the good side of social bots as tools for proper use.

There are more and more researches on social bots. Some studies have studied the influence of social bots on information dissemination in the network environment. The representative ones are the research on the correlation between robots and false news published by Shao et al. [1] in Nature Communication and the research on the role of social bots during the independence referendum in Catalonia, Spain published by Stella et al. [2] in PNAS.

Some researchers have proposed some detection algorithms for malicious social bots. Liu et al. [3] analyzed community structure and used community similarity to separate malicious social bots from humans. Mehrotra et al. [4] proposed an approach to detect fake followers using centrality measures of the nodes in the graph. Moreover, Davis et al. [5] used a feature-based detection method. They evaluated the similarity between a Twitter account and the known characteristics of social bots by more than one thousand features. Wang et al. [6] explored the feasibility of a crowdsourced Sybil detection system for OSNs.

In recent years, intelligent social bot technology is used for carrying out controlled experiments in real social networks and media environment, which is a good application. Murthy et al. [7] tracked and measured the impact of bots on public opinion related to political events by putting social bot accounts on Twitter during the 2015 British election. Monsted et al. [8] deployed social bots on Twitter to analyze label propagation, and verified the effectiveness of different information dissemination models. Min et al. [9] carried out the first controlled experiment by social bots on Weibo in China. In January 2020, Ledford et al. [10] published a paper in Nature in which clearly pointed out how to use social bots to research. Meanwhile, they proposed that combat existing recommendation algorithms and malicious bots was a frontier problem in the field of Internet information dissemination.

Thus it can be seen that social bots have become an important role on the Internet. This chapter summarizes some related knowledge of social bots. Section 11.2 of this chapter introduces the definition of social bots; Sect. 11.3 describes the use and influence of social bots; Sect. 11.4 introduces some related technologies; Sect. 11.5 summarizes some popular malicious bots detection algorithms; Sect. 11.6 is the application of social bots in the controlled experiment; finally, it puts forward the prospect and some existing problems of social bots in experiment.

11.2 Definition of Social Bots

So what is social bot? Although social bot has been widely discussed and studied, there is still no accepted definition so far. Journalists and reports have given their own understanding. There are quite different and may even contradict to some extent among definition. Some definitions may focus on the technology of social bots, while others may focus on what bots do.

Woolley et al. [11] define social bots as a particular type of automated software agent written to gather information, make decisions, and both interact with and imitate real users online. The definition is similar to "A social bot is a computer algorithm that automatically produces content and interacts with humans on social media, trying to emulate and possibly alter their behavior" given by Ferrara [12].

Howard et al. [13] state that social bots are able to rapidly deploy messages, replicate themselves, and pass as human users whatever their uses. They can perform legitimate tasks like delivering news and information, or undertake malicious activities like spamming, harassment and hate speech.

Weedon et al. [14] highlight that automation is a key symbol of social bots. Hwang et al. [15] define that social bots are programs that operate autonomously on social networking sites while Wagner et al. [16] think social bots are automatic or semi-automatic computer programs that mimic humans and human behavior in online social networks.

Grimme et al. [17] give social bot a more detailed definition as a high concept. Their definition covers five aspects: (1) fully automated as well as partly human-steered bot action; (2) autonomous action (agent-like); (3) an orientation toward a goal, (4) multiple modes of communication; (5) a wider ecosystem (all online media). Then they list several examples of social bots, such as chat bot-"a software system, which can interact or chat with a human user in natural language such as English" [18], spam bot, political bots-"spread political content or participate political discussions" [11], mobile phone assistants and so on. In this chapter, we adopt the definition given by Grimme.

11.3 Application and Influence of Social Bots

Social bots are complex. They act like a human, but think like a bot [12]. Actually, they were born to serve humanity. They can do many things on social networks. However, the emergence of any technology may be accompanied by abuse and social bots are no exception.

11.3.1 Application

Although social bots aim to provide services, the emergence of technology is always accompanied by abuse. Nowadays, although some social bots are used to serve society, most of them have been used to carry out malicious acts. In this chapter, we divide social bots applications into five categories.

(1) Dissemination of information: it aims to automatically publish the latest news, blogs, emergencies, or collect a lot of content (such as weather updates, constellation luck, etc.) for human consumption and use.

(2) Scientific experimental tools: social bots can be used as tools for scientific experiments on social networks. This content will be detailed in Sect. 11.6.
(3) Advertising marketing: in a short period of time, social bots can publish a large number of almost the same content to achieve advertising marketing.
(4) Phishing: the malicious link is embedded in the information to cheat human users and steal privacy.
(5) Public opinion guidance: in the process of forming various opinions on social networks, bots aim to guide public opinion and change the formation of public opinions.

11.3.2 Influence

Some social bots that have been maliciously designed, which are used to deliberately do "bad things". So the impact we are discussing here is negative. Social bots can be used to intervene in politics, deliberately guide public opinion in the election process and influence the political election results. Stella et al. [2] analyzed the influence of social bots on the Catalan referendum for independence on October 1, 2017. By using nearly 4 million Twitter posts generated by almost 1 million users, they quantified the structural and emotional roles played by social bots. They claimed that bots increased exposure to negative and inflammatory content in online social systems. A framework to detect a potentially dangerous behavior promoted by bots was proposed. They found that humans and bots had similar time behavior patterns by analyzing the observed social interactions in mentions, replies and retweets. And the sentiment analysis revealed that social bots had almost no emotional deviation when forwarding robot messages, but it had obvious positive or negative emotional trends when forwarding human messages. The result showed that bots choose influential humans, assaulting Independentists with violent content, making Independentists' narratives negative. Then it would aggravate social conflict online.

Ferrara et al. [19] analyzed the influence of social bots on the MacronLeaks disinformation campaign before the 2017 French presidential election. They collected a massive Twitter dataset of nearly 17 million posts that occurred between April 27 and May 7, 2017. They combined machine learning with cognitive behavioral modeling techniques to separate humans from bots and analyze the two groups. They thought the cause of the failure of the campaign was that the users who engaged with MacronLeaks are mostly foreigners with a preexisting interest in alt-right topics and alternative news media, rather than French users with diverse political views. Bessi et al. [20] studied how the social bots influenced political discussion around the 2016 U.S. Presidential election. They discovered about 20% of the entire conversation might not be generated by humans. They analyzed political partisanships for humans and bots and the impact mechanism of the degree of network embeddedness of the bots.

It may also be used to spread some false news, resulting in the widespread of false news. Shao et al. [1] analyzed the spread of low-credibility content by social bot. By collecting and analyzing 14 million messages including 400 thousand articles on Twitter for 10 months during 2016 and 2017, they found that social bots played a disproportionate role in spreading articles from low-credibility sources. At the beginning of spreading, bots exaggerated the content. Meanwhile, bots attracted followers by replies and mentions. And humans could easily believe and reshare the content.

Social bot can be used to steal personal privacy illegally. Boshmaf et al. [21] designed and analyzed a social botnet to evaluate how vulnerable online social networks were to an attack by social bots. They put a group of coordinated programmable social bots on Facebook for 8 weeks and collected data about human behavior in response. The results showed that social bots could attack successfully with a rate of up to 80% and more private data might be exposed because of user profile settings.

Besides, there are still many dangers of social bots, such as manipulating the stock market, being a tool to destroy a person or a company's reputation, disrupting the social network environment, spreading negative emotions on online social networks and so on.

Many social bots are used to destroy the network environment and affect social stability. But as long as they are used properly, they can still provide some efficient and intelligent services and promote social progress, such as search news, automatic reply, intelligent chat and so on. They can also be used for scientific research, and nowadays many researchers have discovered the benefits of social bots as scientific research tools. Their intelligence can replace manpower, even more efficient and flexible. For example, in recent years, some studies have begun to use social bots to conduct controlled experiments, explore the rules of human-computer interaction and the structure and characteristics of social networks, then explore new technologies of public opinion intervention and guidance. This part will be carried out in Sect. 11.6.

11.4 Development Technology of Social Bots

To deploy social bots on the network, first of all, Internet access technology is needed to enable bots to move freely on the network. Then we need to use artificial intelligence technology, so that social bots can realize and simulate human behavior. At the same time, it also needs the knowledge of network science, so that social bots can master the network structure, and "socialize" faster and more directly to achieve the goal. The following summarizes some mainstream technologies.

11.4.1   Internet Access Technology

Internet access refers to connecting computers, mobile devices and other electronic devices to the Internet, so as to realize the information transmission between devices, so that users can use various services provided on the Internet.

11.4.1.1   PC Side

- **HttpClient** HttpClient is a client programming toolkit that supports Http protocol, and it supports the latest version and suggestion of Http protocol. HttpClient has been used in many projects, such as Cactus and HtmlUnit, two famous open source projects on Apache Jakarta.

11.4.1.2   Browser-based Access

- **Selenium WebDriver** WebDriver, also known as Selenium2, is the most important component in the Selenium suite. WebDriver directly invokes the browser through the local interface of the browser automation and performs some automatic operations on the browser, such as opening a specified webpage, obtaining the source code of the current webpage, clicking a certain location in the webpage, simulated browsing, etc. It can automatically simulate human web browsing behavior and avoid JavaScript anti-climbing checks on some websites. And it supports most of the commonly used programming languages, such as Java, Python, etc., which is conducive to multi-platform and multi-thread development. At the same time, you can use a no-interface browser to save machine resources.
- **Puppeteer** Puppeteer is a Node library that controls Chrome or Chromium through the DevTools Protocol. The puppeteer can generate page screenshots or PDFs, grab SPA, generate pre-rendered content, submit the form automatically, etc.
- **Chrome plugins** Chrome plugins can automatically control chrome to make network requests. A chrome plugin is a software developed with web technology to enhance browser functions. It is a .crx suffix compressed package composed of resources such as HTML, CSS, JS, and pictures. The chrome plugin provides many useful APIs, including window control, label control, network request control, various event monitoring, complete communication mechanisms, etc. The chrome plugin can avoid JavaScript's anti-crawling technology and modify the attribute value of WebDriver when chrome loads web pages.

11.4.1.3 Mobile

- **ADB(Android Debug Bridge)** ADB is an Android tool that can be used to connect to an emulator or actual mobile device. ADB can monitor all connected devices (including simulators). At the same time, many commands are provided to control the device.
- **Appium** Appium is an open source and cross-platform test automation framework. Like selenium webdriver, it is also based on the Http protocol and packaged with Node.js. It processes HTTP requests in the same way as Selenium Webdriver; that is, the server receives the Http requests that follow the JSON protocol sent by the client. Appium supports application automation across various platforms such as iOS, android, and windows. Each platform is supported by one or more "drivers" that know how to automate a particular platform.
- **Macaca** Macaca is an open-source automation test solution. It is cross-platform. The application scenario supports the mainstream mobile technology platforms OS, Android, the hybrid runtime Webview of the two platforms, and also supports the previous desktop browsers.

11.4.2 Artificial Intelligence Foundation

Social bots also involve natural language processing, a branch of artificial intelligence. Natural language processing is the study of mathematical and computational modeling of various aspects of language and the development of a wide range of systems [22]. Social bots can interact with humans using natural language processing. Text classification is a branch of natural language processing. In the running process of social bots, text classification is needed, including topic recognition, emotion analysis, intention recognition, etc.

11.4.3 Network Science Theory

Each social network has a unique structure and characteristics. Since robots act on the network, analysis of network structure and transmission is also essential. The knowledge of this part can refer to the book *"Network Science"*.

11.5 Social Bots Detection

Social networks have a large user base. More and more people begin to use various social platforms to integrate their daily lives with social platforms closely. Subsequently, the popularity of social networks also leads to the rise of social bots.

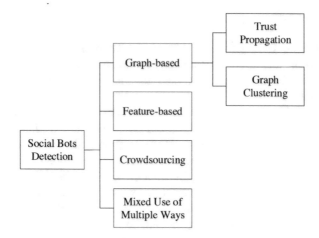

Fig. 11.1 Taxonomy for social bots detection methods

Part of bots steal the private information of users, send spam, spread fake news, and even manipulate social media discourse, which has already endangered the security of cyberspace. Therefore, more and more researchers are engaged in studying how to distinguish social bots from legitimate users in social networks. Generally, the methods are divided into four categories [12]: graph-based detection, feature-based detection, crowdsourcing detection, and mixed use of multiple ways, as shown in Fig. 11.1.

11.5.1 Graph-based Detection Method

The graph-based method involves using the structure of the graph formed by users to detect malicious social bots. Mainstream graph-based algorithms are typically based on the assumption that malicious users may be difficult to connect with a large number of legitimate users in social networks, and thus tend to form their own communities [12]. Based on this assumption, two types of graph-based methods are typically proposed, which are the trust propagation method and the graph clustering method [23].

The general process of trust propagation method is to identify some trusted seed nodes firstly, and then spread the trust according to the connection between the trusted nodes and the unknown nodes. For example, SybilGuard [24], Gatekeeper [25], SybilLimit [26], SybilRank [27], and SybilRadar [28] use random walk technique to propagate trust outwards.

The graph clustering method focuses on the detection of communities to identify malicious group. Liu et al. [3] proposed a community-based method, which contained two major steps. They first adopted the BIGCLAM [29] community detection algorithm to recognize communities and assigned an initial label for each user based on the features exhibited by users. Secondly, they iteratively refined a user's label by combining his/her initial label and his/her community friends.

11.5.2 Feature-based Detection Method

The feature-based method utilizes various account features to train classifiers to detect malicious accounts. Mehrotra et al. [4] proposed an approach to detect fake followers using only network measures of the nodes in the graph. In their work, six network centrality measures were used as features (including Betweenness Centrality, Eigenvector Centrality, Indegree Centrality, Outdegree Centrality, Katz Centrality, and Load Centrality). They applied three classifiers, Artificial Neural Networks, Decision Tree as well as Random Forest, and get the highest accuracy of 95%, with the precision of 88.99%, and recall of 100%.

In contrast to Mehrotra's work, John et al. [30] analyzed the semantic feature of tweet sentiment of legitimate and bot accounts. They proposed SentiBot, a framework for detecting social bots. SentiBot makes use of four categories of features: tweet syntax, tweet semantics, user behavior, and user neighborhood. The results suggest that sentiment plays a significant role in the identification of the bots.

BotOrNot [5] is a feature-based social bot detection architecture, which is a publicly-available service for Twitter and published in 2014. The system leverages more than one thousand features and groups them into six classes which are summarized in Table 11.1. BotOrNot uses Random Forest as the classifier. They collect a dataset of 15k manually verified social bots and 16k legitimate accounts and use the dataset consisting of more than 5.6 million tweets to train the models. After ten-fold cross-validation, it achieves a performance of 0.95 AUC.

Table 11.1 Classes of Features Employed by Feature-Based System for detecting social bots

Class	Description	Feature examples
Network	Network features include various dimensions of information diffusion patterns	Degree distribution, clustering coefficient, centrality measures
User	User features are meta-data related to an account	language, geographic locations, account creation time
Friends	Friend features involve descriptive statistics relative to the social contacts of an account	the distributions of the numbers of followers or followings
Temporal	Timing features reflect temporal patterns of content generation and consumption	Average number of postings per day
Content	Content features are related with the message of the postings of an account	Number of words in a postings, Number of hashtags
Sentiment	Sentiment features are the emotion of the postings of an account	Emotions

11.5.3 Crowdsourcing Detection Method

Crowdsourcing refers to the assignment of tasks to non-specific people in a free and voluntary manner. The crowdsourcing method for social bot detection mainly relies on human intelligence and sensitivity. Wang et al. [6] explored the feasibility of applying human effort (crowdsourcing) like Amazon's Mechanical Turk to detect social bots. They believed that careful users could understand subtle conversational nuances and capture even slight inconsistencies in account profiles and postings. In addition, they designed an experiment that employed both experts and workers to detect social bots simply from the information on their profiles and analyzed the results, which verified their insight. They designed a crowdsourced social bot detection system with two layers. The first layer is filtering layer, in which they can use automated techniques from prior work, such as graph-based detection and feature-based detection, to label the suspicious profiles. The second layer is crowdsourcing layer. The input of this layer is the suspicious profiles recognized by the filtering layer and then a group of workers classifies them as legitimate or fake.

However, there are some drawbacks of the crowdsourcing strategy. First, adopting this method need to hire some experts and large workers, which may lead to a high cost. Second, the exposure of personal information of the users to external crowd workers may cause the issue of privacy. Third, due to the anonymity of crowd workers, some members may accept the task just to get paid, but don't accomplish it seriously and misled the accuracy of detection results.

11.5.4 Mixed Use of Multiple Ways

Advise et al. [31] first proposed the need to build an ever more sophisticated detection system, which combined several detection techniques to effectively detect social bots. The Renren Sybil detector [32, 33] is a hybrid system that combines multiple detection techniques. The system shares the advantages of network graph structure and feature-based detection in some ways. Cao et al. [34] proposed an aggregate behavioral pattern to uncover malicious accounts. Their work was based on the found proposed by Wang et al. [32] that the Http requests from social bots differed from those from legitimate accounts and clustered accounts according to the similarity of their request actions. Secondly, they use pre-labeled accounts to uncovers the clusters of malicious accounts.

11.6 Social Bots and Social Network Control Experiment

At present, the research on online social networks mainly relies on the observation method, which means that researchers can not intervene in the research object, but can only process and analyze the data passively obtained. However, large open

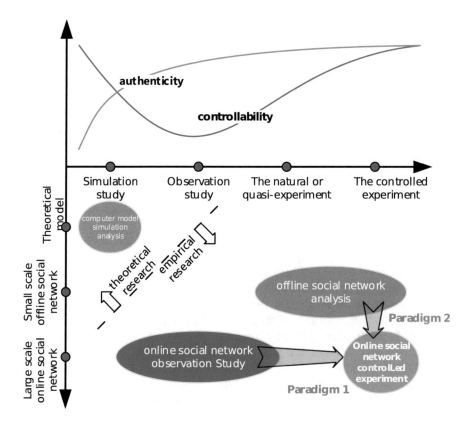

Fig. 11.2 Comparison of social network analyzing methods

data contains a lot of noise and the risk of privacy leakage. In order to overcome the shortcomings of purely observational research, some researchers skillfully use natural experiments or quasi experiments to make comparative analyses and causal inference from available data sets. However, uncontrolled methods limit the research. The comparison of social network analysis methods is shown in Fig. 11.2. Recently, controlled experiments on social networks have deepened our understanding of information sharing and dissemination. Moreover, compared with the big data method, the controlled experiment can effectively control the influencing factors, and the sample set is relatively small [35]. Social bot is being used in controlled experiment research, which may be a new application direction of social bot. With the continuous development of artificial intelligence technology and the rise of big data and social computing technology, social bots can more accurately simulate part of human behavior, which provides a new tool for online social network controlled experimental research.

11.6.1 Online Social Network Controlled Experiment

Controlled experiment refers to the scientific test under controlled conditions, that is, in each experiment, only one or a few factors change, while other factors remain unchanged. Compared with the observation data analysis, the research objects in the controlled experiment are controlled by the experimental designer. Therefore, the causal relationship between the controlled factors and the experimental results can be explored through a small number of factors. Due to the high requirements of controlled experiments, the complex online social network environment, the large number of users and the wide distribution, the controlled experiment method is not common in the research of large-scale online social networks. However, almost all the research results of controlled experiments have a high theoretical level and practical value.

Currently, there are two main modes of online social network controlled experiments (as shown in Fig. 11.2). The first mode can be called the "embedded" mode, which presents the evolution process from the theoretical analysis of complexity science and big data observation research to the controlled experiment. It inherits the abstract model and simulation research ideas in complexity theory analysis and the requirements of real-time and authenticity of big data observation research. The embedded mode is usually directly tested on real large-scale social networks (such as Facebook and Twitter). The scale of the experiment is huge, usually involving tens of thousands or even tens of millions of users. For example, in the experiment of Robert et al. [36], a total of about 61 million Facebook users were involved. Due to the large scale, the control measures taken by this model are usually relatively simple, with only a small amount of intervention in user operations, information and interfaces. At the same time, this model usually requires the cooperation or support of relevant social network service providers. The embedded model takes the real online social network service as the research object, and embeds the experimental intervention into the real network for scientific research.

Another model can be called the "platform" model, which presents an evolution from traditional offline social networks to online social networks. The scale of this model is between traditional offline network experiments and the "embedded" model experiments, usually on the order of a thousand people. At the same time, this model uses Internet technology to create its own experimental social network or application, and does not rely on real social network service providers. Therefore, the platform model is a model that transfers traditional offline controlled experiments to online platform execution. It makes full use of the Internet and computer technology to expand the scale of offline experiments, while being able to control and simulate more complex impact factors. It can simulate the operation and characteristics of some online social networks, and is more convenient for the structural evolution of social networks and psychological experimental research.

In online social network controlled experimental research, the most widely used experimental methods and tools mainly include email services, self developed of social networks/applications, social network customization, and crowdsourcing services. In social network research, the earliest Internet tool to be used is the email service. Classical social network theories such as the small world all rely on email. Self-developed social network application refers to the research on controlled experiments conducted by researchers who independently develop small online experimental social network services or embedded applications developed by using official application interfaces of real social network service providers (such as microdisks on microblog platform). For example, Deters et al. [37] let participants add a Facebook account as their friend, so that they can dynamically obtain all kinds of information authorized by users or regulate user behavior through the account, and carry out analysis and research. Social network customization is led by online social network providers, which directly modifies some functions and displays of existing social network service platforms (such as Microblog, Wechat, Twitter, Facebook, etc.), so as to control the information acquisition or behavior of users. Using social network customization to perform controlled experiments is the most natural method, but at the same time, due to the limitation of service providers, it is also a method with the highest access threshold. For example, during the 2010 U.S. Congress election, Robert et al. [36] conducted a randomized controlled experiment to control whether users could see the voting dynamics of their friends through a customized Facebook interface, and verified the great potential value of online social networks in political mobilization. Crowdsourcing refers to the practice that organizations allocate the tasks which are previously performed by fixed personnel to non-specific mass network users in a free and voluntary manner. The current crowdsourcing platform allows researchers to customize the specific experimental task system by using Internet technology, and open it to the public volunteers for cooperation. For example, Li et al. [38] conducted a social dilemma experiment on the crowdsourcing platform Mturk, and they tested the effects of two networks reciprocity mechanisms and costly punishment mechanisms from the perspective of controlled experiments. The following is a comparison table of these patterns and methods (Table 11.2).

The development of artificial intelligence technology provides a new opportunity for the experimental research of online social network. Social bot based on natural language processing technology has been applied to controlled experiments.

11.6.2   Application of Social Bots in Controlled Experiment

Social bots provide a brand-new tool and model for online social network controlled experimental research, and a new perspective for social network analysis. This model uses real social networks as the experimental environment, using simulation software robots with intelligent behaviors as the subject of the experiment, taking into account the flexibility and controllability of pure numerical simulation analysis, the authenticity of real social network services, and the relatively low standard entry barriers and experimental costs.

Table 11.2 Schema and tool of control experiments in online social network

Experimental methods and tools	Experimental mode	Merits	Demerits
Emails	Platform mode	Low entry threshold; Reflect the real network	Low controllability; Small scope of application; Lack of features of modern social network services
Self developed social networks Applications	Platform/ Embedded mode	High controllability; Reflect part of the real social network	Difficult to develop; High cost
Crowdsourcing service	Platform mode	Good inheritance; Be able to use laboratory scale tools directly; High controllability	Small scope of application; Low social relations and content authenticity
Social networking customization	Embedded mode	Reflect the real network; Wide range of application	High entry threshold; High legal and moral risks

Mønsted et al. [8] deployed social bots on Twitter to realize tag propagation, and verified the effectiveness of different information propagation models. They proposed two Bayesian statistical models describing simple and complex contagion dynamics, and tested the competing hypotheses. Their experiment showed that the complex infection model could describe the observed information diffusion behavior more accurately than the simple infection model. Munger, a political scientist at Pennsylvania State University, using social bots with different user attributes and number of fans to attack accounts with racist comments [39], aimed at reducing white racist speech. It is found that accounts attacked by bots Who have a profile picture of a white man with a large number of fans will significantly reduce their racist comments. The experiment expanded from the laboratory environment to the real network environment, using objective, behavioral results measurement and two consecutive months of data collection period, which is a great progress in behavioral research. Murthy et al. [7] artificially put social bots on Twitter to track and measured the effect of the robot on public opinion related to political events (2015 British election), and measure the degree value of the bots in promoting the formation of topic network and spreading specific tag information. The result shows that the effect of robot is not significant. Chen et al.[40] deployed neutral social bots on Twitter to probe biases that may emerge from interactions between users, platform mechanisms, and manipulation by inauthentic actors. They found evidence of bias affecting the news and information to which U.S. Twitter users were likely to be exposed, depending on their own political alignment.

Min et al. [9] deployed 128 microblog social bots in Weibo (the most famous twitter-like service in China, with 430 million active users per month). By analyzing the data generated by those bots, they clarified the pathway from selective exposure to polarization, especially the structure and function of filter bubbles. This method

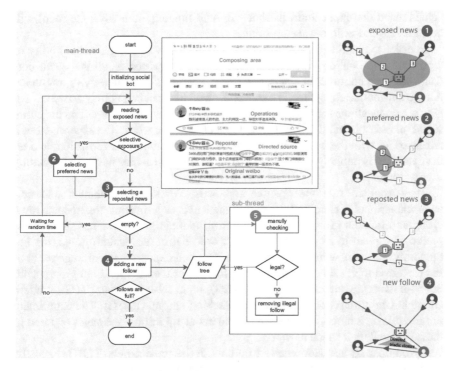

Fig. 11.3 Design of neutral social bot. Based on the operating interface of Weibo (middle), the authors use the flowchart (left) and the schematic diagram (right) to illustrate the main workflow of the social robot, including the automated process (1–4) and the manual assistant process (5)

is limited to the use of data generated by these bots, without any data about actual personnel. Therefore, this is a privacy protection method.

The robot's workflow including five steps is shown in Fig. 11.3. (1) Initially, each bot is assigned two or three default followings, and most of these followings publish or republish messages consistent with the bot's topic. (2) The bot will wake up from the idle state at regular intervals. After the bot is awakened, it can view the latest news published or republished by its followings. In this process, all messages are re-ranked according to the descending order of release time, so as to eliminate the influence of algorithm ranking and recommendation system on information exposure. (3) After viewing the public messages, the bot only selects the messages consistent with the topic. In this step, they first run FastText text classification algorithm to obtain preliminary classification results and then verifies by at least experimenters to ensure the accuracy. Although manual supervision is required, the algorithm can filter out a large number of inconsistent messages. (4) If there are reposted messages in selected messages, according to directed triadic closure, according to directed triadic closure, the bot will randomly select a republished message and follow the direct source of the republished message. (5) If the following values reach the upper limit, the bot will stop running; otherwise, the robot will become idle and wait to wake up again.

In order to avoid legal and moral hazards, the bot will not generate, modify or repost any information in the experiment. In all the processes, all the social bots deployed have no direct interaction with existing users and can only follow them. All the data collected is directly exposed to them and can be publicly accessed. All network connections are restricted to the bot and its direct followings, and the entire social relationship of all followings is not used. In other words, the experiment will not interfere with the user's behavior or the dissemination of information, and all the data used is publicly visible. Therefore, this method avoids the risk of violating user privacy.

They adopted two most active topics in Weibo: entertainment and sci-tech(science and technology), and designed two experimental treatments: topic group and random group for each topic. In the topic group, the bot chooses the preferred content to expand its social network, but in the random group, the bot randomly selects new information sources from all exposed content, regardless of the preferred topic. As a result, they set four experimental social bot groups with 30–34 bots: topic entertainment group (EG), topic sci-tech group (STG), random entertainment group (REG), and random sci-tech group (RSTG). The experiment ran for at least 2 months and they recorded about 1.3 million messages exposed to these bots and their social networks.

By comparing the data received by bots in the topic group(EG,STG) and the random group(REG,RSTG) respectively, they found that the link between selective exposure and information polarization. Figure 11.4 showed that the entertainment topic is more polarized than the sci-tech topic initially. Compared to the initial state, they were more concerned about the diversity of messages consumed by social bots after their social networks have formed. At the end of the bots, the preferred topic ratio of REG decreased more than the preferred topic ratio of EG, which showed that selective exposure was important, but not enough to lead to polarization. On the other hand, the preferred topic ratio of RSTG and STG was similar, both greatly less

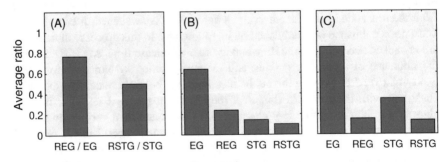

Fig. 11.4 The polarization of exposed content and followings. (**a**) The preferred topic ratio in the initial state (R^0). Because the topic group and the random group have the same initial followings, they also have the same R^0 for each topic. (**b**) The preferred topic ratio in the final state R^1. (**c**) Proportion of followings with the same preferred topic (P) in four treatments

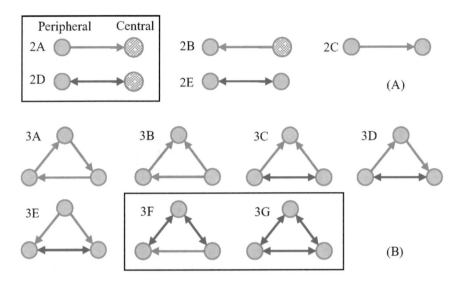

Fig. 11.5 Motifs of personal social networks, distinguish between the central and the peripheral node

than the initial ratio. The difference between the two themes suggested that whether selective exposure led to polarization depends on the topic.

They analyzed all possible motifs between two nodes (i.e., followings of bots) in the networks and there were five possible configurations by considering the centrality of nodes within degree (i.e., the number of followers of nodes). For two nodes of a directed arc, they distinguished two nodes into a peripheral node and a central node if the difference between centralities was larger than a threshold T = 10, such as motif 2A and 2B in Fig. 11.5. Figure 11.6a showed that EG networks contained more non-reciprocal arcs from peripheral nodes to central nodes, whereas STG networks had more reciprocal arcs between peripheral and central nodes. Second, Fig. 11.6b showed that STG networks contained more closed triangles with at least two reciprocal arcs.

Finally, they visualized the evolution of personal social networks. As shown in Fig. 11.7, the STG networks still had a high connection density after removing all the non-reciprocal arcs, while EG networks were the opposite, with the sparse connection. As a result, an STG network was typically a bidirectional clustering structure, while an EG network was typically a unidirectional star-like structure with a few high-degree nodes. The star-like social structure, in which the central nodes played only the role of the information source, and rarely received information from other nodes. The nodes of the bidirectional clustering structure can efficiently exchange information and realize a complementary effect with their friends with distinct preferences, which can promote the diversity of information. The two different structures formed by the EG networks and STG networks led to the different polarization levels.

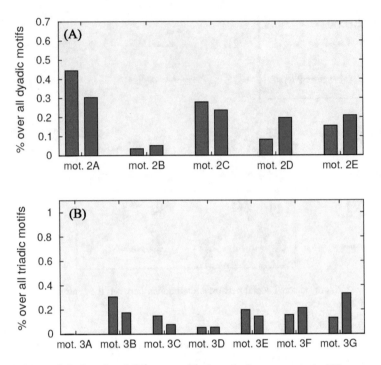

Fig. 11.6 The statistical results of different motifs, the red color represents the EG group and the blue color represents the STG group

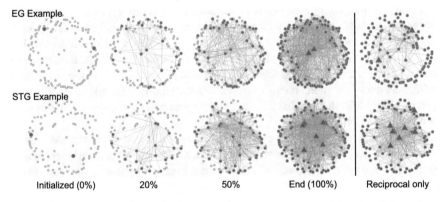

Fig. 11.7 Visualization of the evolution of personal social networks. In this demonstration, Social bot 05 in EG (EG05) and bot 02 in STG (STG02) display a similar growth process from initialized two followings to the end of running. However, STG02 obtains more reciprocal edges than EG05. The triangles represent the nodes with high in-degree, that is, the corresponding users have a large number of followers. The visualization is based on the radial layout with in-degree centrality; thus the node with higher in-degree is closer to the centre of the plot

It can be seen that social bots are a good tool for the development of network control experiments, and at the same time, they can also develop the use of bots in a positive way.

11.6.3 Problems in Controlled Experiments by Social Bots

11.6.3.1 High Technical Threshold

Using social bots to do controlled experiments has high technical requirements. Bots need to make social bots that adapt to different social network structures. For example, the information interaction between Twitter and Weibo is quite different. Social bots on Twitter cannot be applied directly to Weibo, and vice versa. At the same time, all software platforms will be updated, and bots need to make real-time changes. Because of the universality of social bots, the current network needs human-computer verification. Complex text and graphic captcha make the bot unable to run smoothly on the network. Bots are not real people after all, so it still needs a high technical threshold to make bots run on the network like human beings.

11.6.3.2 Legal and Moral Issues

Using social bots for controlled experiments may lead to legal and moral problems. When social bots run on the Internet, it is likely to interfere with the user's behavior, and even produce a series of problems, such as public opinion guidance, invasion of privacy and so on. Scientific research should not infringe on users' privacy, or even interfere in national political activities. When doing experiments, researchers need to pay attention to that the process is only used for scientific research, and cannot cause improper control and intervention, and cause social disputes.

11.7 Conclusion

Nowadays, the negative effects of social bots far outweigh the positive ones. When it comes to social bots, people instinctively associate them with a series of bad words, such as "water army" and "spray". They are influencing and changing the direction of public opinion, political election and so on. Online networks are full of all kinds of malicious social bots. Therefore, it is necessary to detect malicious bots. Section 11.5 summarizes the detection methods of malicious robots from four categories. At the same time, we can make some scientific innovations according to the flexibility and effectiveness of social bots. SEC 6 focuses on the application of social bots in controlled experiments, which provides some inspiration for the future development of social bots.

References

1. Shao, C., Ciampaglia, G.L., Varol, O., Yang, K.-C., Flammini, A., Menczer, F.: The spread of low-credibility content by social bots. Nature Commun. **9**(1), 1–9 (2018)
2. Stella, M., Ferrara, E., De Domenico, M.: Bots increase exposure to negative and inflammatory content in online social systems. Proc. Natl. Acad. Sci. **115**(49), 12435–12440 (2018)
3. Liu, D., Mei, B., Chen, J., Lu, Z., Du, X.: Community based spammer detection in social networks. In: International Conference on Web-Age Information Management, pp. 554–558 (Springer, Berlin, 2015)
4. Mehrotra, A., Sarreddy, M., Singh, S.: Detection of fake twitter followers using graph centrality measures. In: Proceedings of the 2016 2nd International Conference on Contemporary Computing and Informatics (IC3I), pp. 499–504 (IEEE, New York, 2016)
5. Davis, C.A., Varol, O., Ferrara, E., Flammini, A., Menczer, F.: Botornot: A system to evaluate social bots. In: Proceedings of the 25th International Conference Companion on World Wide Web, pp. 273–274 (2016)
6. Wang, G., Mohanlal, M., Wilson, C., Wang, X., Metzger, M., Zheng, H., Zhao, B.Y.: Social turing tests: Crowdsourcing sybil detection. arXiv preprint arXiv:1205.3856 (2012)
7. Murthy, D., Powell, A.B., Tinati, R., Anstead, N., Carr, L., Halford, S.J., Weal, M.: Automation, algorithms, and politicsl bots and political influence: a sociotechnical investigation of social network capital. Int. J. Commun. **10**, 20 (2016)
8. Mønsted, B., Sapieżyński, P., Ferrara, E., Lehmann, S.: Evidence of complex contagion of information in social media: an experiment using twitter bots. PloS One **12**(9), e0184148 (2017)
9. Min, Y., Jiang, T., Jin, C., Li, Q., Jin, X.: Endogenetic structure of filter bubble in social networks. R. Soc. Open Sci. **6**(11), 190868 (2019)
10. Ledford, H.: Social scientists battle bots to glean insights from online chatter. Nature **578**(7793), 17–17 (2020)
11. Woolley, S.C.: Automating power: social bot interference in global politics. First Monday **21**(4) (2016)
12. Ferrara, E., Varol, O., Davis, C., Menczer, F., Flammini, A.: The rise of social bots. Commun. ACM **59**(7), 96–104 (2016)
13. Howard, P.N., Kollanyi, B., Woolley, S.: Bots and automation over twitter during the US election. In: Computational Propaganda Project: Working Paper Series (2016)
14. Weedon, J., Nuland, W., Stamos, A.: Information operations and facebook. In: Retrieved from Facebook. https://fbnewsroomus.files.wordpress.com/2017/04/facebook-and-information-operations-v1.pdf (2017)
15. Hwang, T., Pearce, I., Nanis, M.: Socialbots: voices from the fronts. Interactions **19**(2), 38–45 (2012)
16. Wagner, C., Mitter, S., Körner, C., Strohmaier, M.: When social bots attack: Modeling susceptibility of users in online social networks. In: # MSM, pp. 41–48 (2012)
17. Grimme, C., Preuss, M., Adam, L., Trautmann, H.: Social bots: human-like by means of human control? Big Data **5**(4), 279–293 (2017)
18. Shawar, B.A., Atwell, E.: Chatbots: are they really useful? In: Ldv Forum, vol. 22, pp. 29–49 (2007)
19. Ferrara, E.: Disinformation and social bot operations in the run up to the 2017 French presidential election. arXiv preprint arXiv:1707.00086 (2017)
20. Bessi, A., Ferrara, E.: Social bots distort the 2016 US presidential election online discussion. First Monday **21**(11-7), 14 (2016)

21. Boshmaf, Y., Muslukhov, I., Beznosov, K., Ripeanu, M.: Design and analysis of a social botnet. Comput. Networks **57**(2), 556–578 (2013)
22. Reshamwala, A., Mishra, D., Pawar, P.: Review on natural language processing. IRACST Eng. Sci. Technol. An Int. J. (ESTIJ) **3**(1), 113–116 (2013)
23. Adewole, K.S., Anuar, N.B., Kamsin, A., Varathan, K.D., Razak, S.A.: Malicious accounts: Dark of the social networks. J. Netw. Comput. Appl. **79**, 41–67 (2017)
24. Yu, H., Kaminsky, M., Gibbons, P.B., Flaxman, A.: Sybilguard: defending against sybil attacks via social networks. In: Proceedings of the 2006 Conference on Applications, Technologies, Architectures, and Protocols for Computer Communications, pp. 267–278 (2006)
25. Tran, N., Li, J., Subramanian, L., Chow, S.S.M.: Optimal sybil-resilient node admission control. In: 2011 Proceedings IEEE INFOCOM, pp. 3218–3226 (IEEE, New York, 2011)
26. Yu, H., Gibbons, P.B., Kaminsky, M., Xiao, F.: Sybillimit: a near-optimal social network defense against sybil attacks. In: Proceedings of the 2008 IEEE Symposium on Security and Privacy (SP 2008), pp. 3–17 (IEEE, New York, 2008)
27. Cao, Q., Sirivianos, M., Yang, X., Pregueiro, T.: Aiding the detection of fake accounts in large scale social online services. In: Presented as part of the 9th {USENIX} Symposium on Networked Systems Design and Implementation ({NSDI}12), pp. 197–210 (2012)
28. Mulamba, D., Ray, I., Ray, I.: Sybilradar: a graph-structure based framework for sybil detection in on-line social networks. In: IFIP International Conference on ICT Systems Security and Privacy Protection, pp. 179–193 (Springer, New York, 2016)
29. Yang, J., Leskovec, J.: Overlapping community detection at scale: a nonnegative matrix factorization approach. In: Proceedings of the Sixth ACM International Conference on Web Search and Data Mining, pp. 587–596 (2013)
30. Dickerson, J.P., Kagan, V., Subrahmanian, V.S.: Using sentiment to detect bots on twitter: are humans more opinionated than bots? In: Proceedings of the 2014 IEEE/ACM International Conference on Advances in Social Networks Analysis and Mining (ASONAM 2014), pp. 620–627 (IEEE, New York, 2014)
31. Alvisi, L., Clement, A., Epasto, A., Lattanzi, S., Panconesi, A.: Sok: the evolution of sybil defense via social networks. In: Proceedings of the 2013 IEEE Symposium on Security and Privacy, pp. 382–396 (IEEE, New York, 2013)
32. Wang, G., Konolige, T., Wilson, C., Wang, X., Zheng, H., Zhao, B.Y.: You are how you click: Clickstream analysis for sybil detection. In: Proceedings of the 22nd {USENIX} Security Symposium ({USENIX} Security 13), pp. 241–256 (2013)
33. Yang, Z., Wilson, C., Wang, X., Gao, T., Zhao, B.Y., Dai, Y.: Uncovering social network sybils in the wild. *ACM Trans. Knowl. Discov. Data (TKDD)* **8**(1), 1–29 (2014)
34. Cao, Q., Yang, X., Yu, J., Palow, C.: Uncovering large groups of active malicious accounts in online social networks. In: Proceedings of the 2014 ACM SIGSAC Conference on Computer and Communications Security, pp. 477–488 (2014)
35. C. Jin, Jiang, T., Min, Y., Jin, X., Ge, Y., Chang, J.: Review of control experiments on online social networks. J. Zhejiang Univ. (Science Edition) **47**(1), 1–11 (2020)
36. Bond, R.M., Fariss, C.J., Jones, J.J., Kramer, A.D.I., Marlow, C., Settle, J.E., Fowler, J.H.: A 61-million-person experiment in social influence and political mobilization. Nature **489**(7415), 295–298 (2012)
37. Deters, F.G., Mehl, M.R.: Does posting facebook status updates increase or decrease loneliness? an online social networking experiment. Soc. Psychol. Personal. Sci. **4**(5), 579–586 (2013)
38. Li, X., Jusup, M., Wang, Z., Li, H., Shi, L., Podobnik, B., Stanley, H.E., Havlin, S., Boccaletti, S.: Punishment diminishes the benefits of network reciprocity in social dilemma experiments. Proc. Natl. Acad. Sci. **115**(1), 30–35 (2018)
39. Munger, K.: Tweetment effects on the tweeted: Experimentally reducing racist harassment. Polit. Behav. **39**(3), 629–649 (2017)
40. Chen, W., Pacheco, D., Yang, K.-C., Menczer, F.: Neutral Bots Reveal Political Bias on Social Media (2020)

Printed in the United States
by Baker & Taylor Publisher Services